A
MODERN DAY
YANKEE IN A
CONNECTICUT
COURT

Also by Alan Lightman

Time Travel and Papa Joe's Pipe

A MODERN DAY YANKEE IN A CONNECTICUT COURT

And Other Essays on Science

ALAN LIGHTMAN

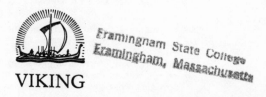

VIKING

VIKING
Viking Penguin Inc., 40 West 23rd Street,
New York, New York 10010, U.S.A.
Penguin Books Ltd, Harmondsworth,
Middlesex, England
Penguin Books Australia Ltd, Ringwood,
Victoria, Australia
Penguin Books Canada Limited, 2801 John Street,
Markham, Ontario, Canada L3R 1B4
Penguin Books (N.Z.) Ltd, 182–190 Wairau Road,
Auckland 10, New Zealand

First published in 1986 by Viking Penguin Inc.
Published simultaneously in Canada

The following essays were previously published, some of them in
different form: "Conversations with Papa Joe" (as three essays);
"Smile"; "E.T. Call Harvard"; "A Flash of Light"; "A Day in
December"; "In His Image"; "A Telegram from Clarence"; "The
Origin of the Universe"; "Tiny Patterns" under the title "Snow";
"Walden" under the title "Walden Pond"; "To Cleave an Atom,"
"A Modest Proposal," and "The Dark Night Sky" in *Science 84,
Science 85,* and *Science 86;* "Rendezvous" under the title "Halley's
Comet" and "Gravitational Waves" under the title "Gravitational"
in *The New Yorker;* and "Elapsed Expectations" in *The New York
Times Magazine.*

"Lost in Space" is reprinted by permission of the *Bulletin of the
Atomic Scientists,* a magazine of science and world affairs. Copyright
© 1984 by the Educational Foundation for Nuclear Science,
Chicago, IL 60637.

LIBRARY OF CONGRESS CATALOGING IN PUBLICATION DATA
Lightman, Alan, 1948–
A modern day yankee in a Connecticut court.
1. Science—Miscellanea. 2. Technology—Miscellanea.
I. Title.
Q173.L724 1986   500   85–41106
ISBN 0-670-81239-0

Printed in the United States of America by
The Book Press, Brattleboro, Vermont
Designed by Helen L. Granger/Levavi & Levavi
Set in Garamond

*The human understanding is no dry light,
but receives an infusion from the will and
affections; whence proceed sciences which
may be called "sciences as one would." For
what a man had rather were true he more
readily believes.*

—FRANCIS BACON
The New Organon (1620)

ACKNOWLEDGMENTS

It gives me pleasure to mention some of the people who encouraged and inspired me as I wrote this second collection of essays. Among my scientific colleagues at many institutions, I especially thank Claude Canizares, Owen Gingerich, Don Lamb, Irwin Shapiro, Kip Thorne, and Alar Toomre, for their continued support and receptiveness to writing about science in human terms. I am grateful to Steven Jay Gould, John McPhee, and Lillian Ross for their encouragement and for what I have learned about writing from their writing. Lucile Burt, Dick Moore, and David Roe have given me keen editorial advice, insight into the human aspects of science, and friendship. To Jim Cracraft, Allen Hammond, David Miller, and George Thompson I am grateful for suggesting the topics of some of the essays. I thank the excellent staff of *Science 86,* where many of these essays first appeared; it has been a pleasure to work with Bonnie Gordon and Laura Ackerman. My colleagues and students at Harvard and the Smithsonian Astrophysical Observatory have provided an invigorating atmosphere for doing science. I thank my literary agent, Jane Gelfman, who expressed her confidence in me. Dan Frank, my editor at Viking, has inspired and guided me with his enthusiasm and intelligence. Finally, I thank my parents, who have remained encouraging as their son the scientist lifted off into the nethers of journalism, and I thank my wife Jean and daughter Elyse, who have opened up my life and given me so much more to write about.

CONTENTS

A FLASH
OF LIGHT

My serious interest in physics began in my freshman year of college. In the dining hall that year, one of the upperclassmen smugly announced that, on the strength of mechanics alone, he could predict where to strike a billiard ball so that it would roll with no sliding. I was mightily impressed and decided this was a subject worth looking into.

Although I didn't realize it at the time, scientists generally divide into two camps, theorists and experimentalists. The abstractionists and the tinkerers. Especially in the physical sciences, the distinction can be spotted straight off. It has since been my observation that, in addition to their skills in the lab, the experimentalists (particularly the males) can fix things around the house, know what's happening under the hood of a car, and have a special appeal to the opposite sex. Theorists stick to their own gifts, like engaging themselves for hours with a mostly blank sheet of paper and discussing chess problems at lunch. Sometime in college, either by genes or by accident, a budding scientist starts drifting one way or the other. From then on, things are pretty much settled.

My path was decided in junior year. For some reason, the physics department had gotten into its head that we students should have a practical knowledge of our subject. To this end, an ungraded electronics workshop, to be tackled in the fall of that year, was strongly encouraged. Most of my colleagues leaped at the opportunity. This was particularly true of those scholars shaky in course work, who could be heard muttering such quips as, "This will separate the men from the boys." (My college was all male in those days.) I had an inkling of trouble, but was not one to tuck in my tail. I signed up.

This electronics workshop was considerably different from the routine laboratory exercises attached to most courses. In the latter, you were always measuring something where you knew damn well what the right answer was. One experiment I remember involved determining the speed of light. The equipment consisted of two mirrors, one of them stationary and the other rotating rapidly. Light making a round trip between the two mirrors would be slightly deflected on its return path by the rotation of the moving mirror, and from the amount of deflection you could deduce the speed of light. Of course, you could also look up the speed of light in any number of books. If your own measured value came out shy of the mark, you could nudge the mirrors and try again. With enough stamina, you eventually got the result you were looking for, at which point everything was carefully recorded, the experiment was declared a success, and you strode out of the laboratory in search of other mountains to climb.

But this electronics project was different. Each of us was provided with a large stock of transistors, capacitors, and so forth, a description of what the final thing had to *do,* and let loose. The stated aim of my gadget, as I recall, was to light up for several seconds when pure tones above middle C were offered to it but to maintain a state of torpor otherwise. (The state of torpor I had no problem with.) To help us get started with the fundamentals, we were given a textbook titled *Basic Electronics for Scientists,* which I immediately recognized as a friendly ally, took with me everywhere for months, and pored over deep into the night, at the expense of my own and my roommate's sleep.

The next couple of months were miserable. I discovered that what worked in the book didn't necessarily work on the lab table, at least under my supervision. In this regard, I lagged far behind most of my classmates. When they looked at the wavy line of an oscilloscope, it spoke to them, and they knew just how to fix up their circuits to get the desired results. I badly wanted my project to succeed. But I did not possess that peculiar knack for making things work. I could write poetry, I could play "Clair de Lune" on the piano, and I loved talking about ideas. But I could not make things work.

One day that term, through some odd misdirection of the postal service, I received in my mail slot a catalog for a home electronics course. Normally, I throw such things out. But to me, at that time, this catalog seemed like a greeting from providence. I took the thing back to my dormitory room, discreetly, and began reading. The front page said something to the effect that, with no prior training or aptitude, you would in six weeks be designing working circuits, mending broken televisions, and presenting yourself to the electronics industry as a force to be reckoned with. There were a few sample diagrams, some pictures of robust-looking devices, and glowing statements by successful graduates. What caught my eye was the provision that, during the course, you could mail in detailed sketches for an electrical project of your own design and be promptly and accurately informed whether the thing would, in fact, work. This last feature was absolutely guaranteed. Mending televisions didn't much interest me, but the chance of securing a foolproof verdict on my floundering electrical enterprise was nothing to be sneered at.

I enrolled without delay in this mail-order electronics course. The cost was $200, and you had to furnish your own parts, of which I had ample supply. My plan was to send in furtively a series of intermediate designs for my college project until one of them received the seal of approval. I could then prance into the physics lab and assemble the contraption in short order. My colleagues, meanwhile, were reporting to the lab daily, laboriously testing out each little step of their assigned projects. I had tried this method and failed. It was a great relief to me that I could now suffer through all the preliminary defeats in private, without humiliating myself in front of the others.

Eventually one of my designs was certified. I spent the last few days before the project deadline calmly at work in the lab, soldering each part into its approved position. My fellow students watched my miraculous progress with the kind of respect that is never verbalized. We were all equals, and I basked in my satisfaction. However, I never had the courage to put the device through a dry run.

The final judgments of the projects were pronounced on a day in December by a highly competent member of the faculty named Professor Pollock. Pollock was a man of few words, but a fair man. He was partially bald, as I remember, wore thick glasses, and usually held his head lowered below eye level. When something you said or did amused him, he would look up briefly and grin, without making the slightest sound or the tiniest movement of his head. Pollock was someone who could make things work. He had large hands. He had built cyclotrons.

On that day in the lab, the various projects and students stood at attention, like dogs and their masters in a dog show. When it came time to put my pet through its paces, I played a note to it—I forget now whether above or below middle C—and it answered with a single, blinding flash of light followed by the unmistakable odor of an electrical fire. The flash going off seemed like a shotgun blast, and I instinctively ducked for cover. It was amazing that no one was hurt. Pollock stood grinning longer than usual.

The following summer Armstrong and Aldrin walked on the moon. As I sat watching them on television in my family house, I had enormous respect for the parts that had worked to get them there: the rocket propellant, the computers, the space suits. And I was filled with admiration for the people behind all that, people good with their hands. Pollock may have been one of them and, undoubtedly, the handy students in my physics class would work on similar things in the future. But it also occurred to me that we theorists were needed to make sure the moon would be there at the same time the astronauts were. There are abstractionists and there are tinkerers, and I was not unhappy to have discovered my lot.

CONVERSATIONS WITH PAPA JOE

The First Evening

An extraordinary thing happened one night last winter. I was relaxing in my study after a long day at the university. As I sat reading, drawing on my great-grandfather's pipe, the old gentleman himself materialized in the curling gray smoke and seated himself in the comfortable wingback across from me. He seemed much less surprised than I and immediately occupied himself with dusting off his suit, as if he'd been on a long journey.

I should explain that I know little of Papa Joe. He came to this country from Hungary about 1880, in his early teens, and started a construction company in Nashville. According to my older relatives, he was not formally educated, but a capable man, with a good head on his shoulders and a strong curiosity about the world. His pipe, a fine old English briar with a solid bowl and a beautiful straight grain, had been tucked away in a drawer for years when my father found it and gave it to me. This was only the second time I had lit it.

After introducing ourselves, we settled into conversation. "I've been looking for that pipe," the old gentleman said, taking a deep whiff of the aromas filling the room.

"It's a wonderful pipe," I agreed. "It's always made me wonder what you were like."

Papa Joe was eager, of course, to learn what had happened in the last sixty years and began asking questions. We talked of how his various descendants had got on in life, the Great Depression, the Second World War, the landing of men on the moon.

All of a sudden, I realized I hadn't offered Papa Joe his own pipe. I wiped off the stem and held it out to him. He reached for it immediately, but then hesitated, and finally pulled back his hand. "What's been has been," he said with a sigh, then walked over to the large bay window behind my desk and stood looking out.

It was one of those crystal nights, with cold, clear skies. Even from my chair near the fire, I could easily make out Orion, with Betelgeuse at the hunter's shoulder and Rigel marking his left foot. Taurus the Bull was close by, glistening through the branches of my maple tree.

"I love the sky at night," said Papa Joe. "Never knew much about the stars, but I always wanted to." He paused, in thought. "I used to tell your father that each star was a firefly."

"I suppose the nights you remember were clearer than this," I said. "Our streetlamps and city lights spoil the view a bit."

He nodded. "But you're not bad off here, on this little street. Not bad off at all. My pipe's found a good home." Just as the old gentleman uttered those words, he started to fade.

"Wait! Wait!" I cried out to him. "I can tell you some things about the sky, if you're interested. I'm actually an astronomer."

At that, my great-grandfather's figure grew firm once again. "I reckon I'll stay awhile then," he said, and returned his gaze to the window. "I trust the sky. Clothes and men change with the styles, but not the stars."

"If you don't look too long, or too far," I said.

"What do you mean?" he asked, turning back toward me.

"You can be pretty sure that each of those stars up there will eventually dim to a cinder or blow itself apart. It's only a matter of time."

Papa Joe had a stricken look on his face, like a man who'd suddenly lost an old friend. I felt a wave of embarrassment. I tried to change the subject, but he wouldn't let me. Instead, he pressed me to explain my remark. I didn't know exactly where to begin, so I put another log on the fire. Papa Joe returned to his chair.

"One thing I have to tell you about modern science," I finally said, "is that it has galloped off into territories far beyond where we can follow with our bodies. What we experience directly with our human senses is only a small fraction of the world around us. But we very badly want to see what our eyes cannot see, and hear what our ears cannot hear. We want to know about places beyond the stars and about happenings before the earth was formed. So we've built enormous machines that dissect the insides of atoms. We've built telescopes that peer out to unimaginable distances and instruments that record colors invisible to the human eye. Our theorists have worked out equations to describe the beginning of time.

"A lot of what we now believe about the world has come to us only by looking at the readings of our instruments and trusting the logic of our calculations. Of all people today, I think scientists have the deepest faith in the unseen world. The greater the scientist, the deeper his faith."

"That's a turn of events," exclaimed my great-grandfather. "I always thought of scientists as fellows who wouldn't forecast rain until they were drenched. This is pleasant news."

"It's a special brand of faith," I continued. "You might say that the scientist sees God as a mathematician. And with some justification, as far as I can tell. As our artificial eyes and ears have revealed each new patch of the invisible tapestry, it looks more and more precise. And our abstract equations and scribblings work remarkably well at predicting the patterns."

"You've whetted my appetite, young man. But please mind, I'm not much with philosophy. I like to have solid proof for what I believe."

"Do you believe the earth spins on its axis, Papa Joe?"

"Yes."

"What solid proof do you have?" said I. "Do you feel yourself whipped around through space at several hundred miles per hour?"

Papa Joe started to speak, twisted his thick moustache uneasily, and said nothing.

"If you set swinging a long heavy pendulum and watched it carefully, the way Monsieur Foucault first did in the last century, you'd notice that its plane of motion very slowly rotates. That plus some principles of physics prove that the earth turns on its axis. But you'd never catch the tiny effect with your own senses."

The old gentleman chuckled. "All right, I get your point. I'm all ears to what you've learned with your modern devices, if ears are of use anymore. Now tell me about the heavens, misbehaving behind our backs."

"First," I said, "we need to get some idea of the distances. But that's not easy. The prickliest problem in astronomy has been finding the distances to the stars. When we look into the sky, we perceive length and width, but not depth. From our vantage, stars are just white dots on the night sky, like distant ships on a night sea, visible only by their running lights. Some are certainly closer than others, but which ones? How can we measure the size and shape of space itself, stretching all around us? Astronomers puzzled over this problem for thousands of years, knowing that it held the answer to so many other celestial mysteries."

"I'm surprised you can't figure depths with your telescopes."

"Look at any star through the most powerful telescope," I said, "and it will appear as a mere point of light. How do you gauge something like that? And all you've got for reference are other points of light."

"I guess that must mean that the stars are very small, or else very far away," said Papa Joe.

"They're not small," I replied, "but you're right about the distance. If the stars were nearby, then we'd see their locations shift back and forth as the earth moves from one side of the sun to the other, changing our angle of view. In fact, we do see a slight yearly shift in the closest stars and can measure their distances by the amount of the shift. The nearest star is several thousand times farther away than Pluto. But the great majority of stars are so distant that they appear fixed while we go back and forth around the sun."

"Surely," spoke the old gentleman, "the nearer ships must look brighter and the ones farther away must look dimmer. Couldn't their distances be judged in such a way?"

"Aha," I answered, "you're on the right track. But you're assuming that all of the ships carry the same lights on board. Some of the ships, the grander ones, will have stronger beacons, so at a great distance they will appear just as bright as the closer but less luminous ones."

"I should have guessed that the stars, like everything else, would have their own privates and captains," said Papa Joe. "I reckon the first step might be to group the stars by kind somehow, although I can't see how to do it. Then the dimness and brightness trick could be used on stars of the same kind." Papa Joe smiled faintly, as if pleased with his comments.

"That's in fact very close to what Professor Shapley did several decades ago," I said. "Astronomers had noticed that certain stars change brightness in a rhythmic and regular fashion, with some running through their light cycles rapidly and others more slowly. Shapley put these pulsating stars into groups, according to the length of their cycles. Then he used the assumption that every star in the same group was identical, with the same luminosity. For example, every star with a light cycle between ten hours and eleven hours would be in one group; every star with a cycle between eleven and twelve hours would be in another, and so on. With this clever way of identifying what kind of star he was looking at, Shapley could then use the dimness and brightness method to figure its distance. So the pulsating stars became points of reference, at known distances. Find a pulsating star lodged within a group of other stars and you know the distance to them too. Little by little, Shapley began mapping out the heavens and placing many of the points of light at their proper depths, with better accuracy than anyone had managed before. It was immensely tedious work, requiring the scrutiny of thousands of telescopic photographs over time, in order to see which stars changed brightness and how quickly."

"I'm pleased to hear," said Papa Joe, "that your Professor Shapley had to put in some hard work at his job. That makes me believe him all the more. From what you said before, I had the notion that modern scientists simply had to turn on their machines and lie back while new knowledge was cranked out and charted. If you'll pardon me, it sometimes seems that progress breeds laziness. For years, I had a fellow running my stone quarry outside Nashville. Once the telephone lines came in, he started calling me up with every damn fleabite, instead of thinking them out like he used to do. But I've carried us off the point. What did Shapley's labors turn up?"

"For one thing," I answered, "the heavens extend much farther than astronomers previously thought. For another, we're not at the center, any more than our planet is at the center of our solar system. Our sun seems to be casually dropped at the outskirts of an enormous, disk-shaped gathering of stars, called a galaxy, containing every star our eyes can see and a hundred billion more. Before Shapley, astronomers thought our sun was at the center of this galaxy. But the center is far off, in the direction of Sagittarius. The dimensions of the whole thing are fantastic. If our solar system were the size of a dime, then the galaxy would be the size of Tennessee."

My great-grandfather let out a whistle. "I can't imagine anything that large. But being off center could have its advantages," he offered. "It might keep us from getting too stuffed with ourselves. And what's out past the galaxy?"

"Other galaxies, with a lot of mostly empty space in between. As far as our telescopes can see, there are galaxies. Picture yourself gliding through the depths of the universe. You come to a flotilla of stars, all huddled together. That's a galaxy. After you've left the first galaxy far behind, so it's a tiny white patch of fuzz in the dark, you come to another huddling of stars. That's another galaxy. You pass one galaxy after another, some shaped like pinwheels, some like spheres, some like nothing in particular. Then you come to your own galaxy, the Milky Way. You quickly search for your own sun and can hardly find it, a single speck lost in the billions of other specks. The earth is invisible. Then you are gone and your galaxy dwindles behind you, becomes nothing. More galaxies come and go, come and go."

Papa Joe had walked to the window and was looking out at the sky again. He stood there a long while.

"And Professor Shapley," he said softly, "worked it out in an office somewhere, with his photographs and his good head. He sure was small compared to what he was thinking about. That's powerful faith. Powerful faith."

As Papa Joe whispered these last words, his figure grew misty and began to dissolve. I noticed my pipe had gone out. "Don't go," I called out. "There's much more I haven't told you."

"All right. I'll be back tomorrow night," came a wisp of a voice.

"Tomorrow night," I repeated, and then he was gone.

The Second Evening

The next evening, I hurried through dinner and, about eight o'clock, went to my study. I lit up my pipe and drew furiously, filling the room with great clouds of smoke, but nothing happened. Then, when I was starting to feel dizzy, Papa Joe appeared, just as he had the night before. He stood tall and erect for a moment and then sat down across from me, in his chosen chair.

For a while, neither of us spoke. Papa Joe seemed to be enjoying the aromas wafting from the pipe, and I hated to interrupt his pleasure. I'd filled the pipe with my own blend of cavendish and burley, but, as happens with a fine old briar, all the tobaccos he'd ever smoked in it had left their own flavors inside the bowl and were now drifting through the room.

"I'm happy you came back," I finally said. "I hope our conversation last night didn't upset you."

"I must admit, your modern view of the heavens takes some getting used to. It strains me to picture a galaxy and its billions of suns. I have a much easier time picturing a house, with the plans and drawings all in front of me."

"Perhaps that's because you've put up a lot of houses with your own hands. You know how the marks on the drawings will turn into windows and doors."

"Just what I was getting at," said the old gentleman. "It seems that your astronomers want me to imagine an enormous building I've got no way of touching, and neither do they. All they've given me for blueprints are photographs of small white dots, and arguments. The reasoning is sound, I'll agree, but I keep remembering Aesop's astronomer, who walked outside every night looking up at the sky, until one night he fell into a well."

"Don't worry," I said laughing, "we won't have to venture from our chairs. We can continue last night's tour just fine from where we're sitting."

"Good, let's go on. You left off with the galaxies, far apart like little islands in space, except they're not little."

"Now you have to imagine that these galaxies are

flying away from each other at great speed," I said.
"That we learned about fifty years ago from Professor
Hubble, who discovered that galaxies appear redder
than they should be if they were standing still."

"Hold on, you've lost me."

"Let me try and explain with an analogy to sound.
When something making a sound moves, the pitch of
the sound changes. It goes up when the thing's coming
toward you, and down when it's going away. The
faster the speed, the greater the shift in pitch. You've
probably noticed the effect with a passing train. When
it's approaching, the pitch of its whistle rises, and
when it's going away the pitch drops."

The old gentleman nodded. "I know trains pretty
good. The L. and N. ran next to my quarry. Many
times I heard that falling shriek as it passed, but I never
thought much about it."

"Well, the same kind of thing happens with light,"
I went on. "In light, what corresponds to pitch is color.
When a source of light is moving toward you, its color
goes up in frequency, which means it becomes bluer.
When it's moving away, it gets redder. At ordinary
speeds, the change in color is so slight that your eyes
can't see it, but certain very sensitive instruments can.
Hubble had one of these fastened to his telescope
while he was studying galaxies. When he found that
their colors were shifted toward the red, he concluded
the galaxies were traveling away from him in all directions. What's happening, we believe, is that every galaxy is rushing away from every other galaxy, like dots
painted on an expanding balloon. The whole universe
is expanding."

"Hold on, young man. You understand this business of colors much better than I do, but it seems to
me that if the galaxies are flying off, we should see
them move in our telescopes. Shouldn't we?"

"Not if they're very far away," I replied. "Motion at great distance is hard to detect. Galaxies are so far away they seem to be standing still, even in telescopes. Fortunately, we have our spectrometers."

"I'm beginning to feel better and better about being anchored to this chair, with the universe flying apart all around me," said Papa Joe. "I never guessed so much commotion was going on out there."

"You're in with some good company," I replied. "Aristotle convinced everybody the universe was perfectly steady, and people believed him for two thousand years. He had some exhausting arguments, and there wasn't any evidence to the contrary—not until Professor Hubble."

"And if I understand you," said Papa Joe, "you're saying that after all those centuries of peaceful nights under the stars, your modern astronomers have decided that Creation is bursting apart, on the strength of some gadgets looking at little smudges of light through a telescope."

"That's what I'm saying. And I believe it, although I admit it goes against what I see when I look up at the sky."

I got up and took out a pipe cleaner from my desk near the window. My great-grandfather sat working his moustache. "I reckon common sense isn't worth much in this business," he mused.

"It seems to me," I replied, "that common sense is what you learn from personal experience. But we're talking about things that you can't possibly experience, not with your human senses anyway. A good deal of science these days is beyond the senses, and it isn't at all common. The only way to get there is to start with what you're dead sure about, then climb out a bit, standing on solid logic, then climb a little further, inching your way along and making certain each step is firmly supported by the one below. Sometimes you take what you thought was a little step and find yourself hanging in thin air. Then you have to grab on and scramble back a few rungs. One way or another, you eventually get so far up you can't see where you started. That's when you need to have faith."

"I'll bet nothing compares to that feeling of being up in the clouds," said Papa Joe, "with the ground out of sight, and knowing the strength of your ladder. That must be how Shapley felt. And Hubble. I wish I'd been there."

I nodded. "So do I. Those guys had faith—but well-grounded faith, I believe. Take Hubble's, for example. The same spectrometers we point at galaxies we also point at lightbulbs set up in the lab, where we're darn sure whether the lightbulbs are moving or not, and how fast. The theory checks out. So if galaxies aren't flying apart as we think, then the laws of nature in space are different from what they are on the ground. That would be illogical. If one and one make two over here, one and one should make two over there. Or else all science would be in a terrible mess, and scientists would be out of work.

"Let's assume Dr. Hubble was right," I continued, "and the universe is expanding. That means it was smaller and denser in the past."

Papa Joe nodded cautiously, like a man readying himself for the pitch of a used-buggy salesman.

"Then if you mentally go backward in time," I went on, "the galaxies get closer and closer together. Eventually, they touch and merge and become a single mass, which gets denser and denser. Planets and stars lose their boundaries. Atoms get ripped apart and crushed together. Everything gets squeezed closer and closer together. Finally, there comes a definite time in the past when all the matter of the universe is compressed into a single point. Astronomers can estimate that time by measuring how fast the universe is expanding now. It's about ten billion years ago. Ten billion years ago, according to the theory, the universe exploded from a point and was born. Scientists call that beginning the Big Bang."

The old gentleman was busily working his moustache again. Furthermore, he had abandoned the safety of his chair and was pacing the room, narrowly missing the logs piled by the fireplace. "On the strength of some gadgets looking at little smudges of light through a telescope," he muttered. "I used to think *I* had chutzpah."

"It comes with the profession these days," I said. Just then, a church clock struck ten in the distance. Papa Joe produced from his vest pocket a beautiful gold watch, flipped open its cover, and nodded appreciatively. When he saw how taken I was with his watch, he handed it to me to look at more closely. Then he began complaining again about the Big Bang.

"There's something else that adds weight to this ten billion years," I offered. "Stars and planets began forming soon after the universe began, so the earth has to be younger than the universe, but probably not a lot younger. At the beginning of the century, before people had any idea of a Big Bang, some chemists found a way to tell how old the earth is. Special kinds of atoms are continuously changing into other kinds of atoms, in a regular way. For example, uranium atoms change into lead atoms. If you start off with a rock of pure uranium, after a certain number of years half of it will be lead. After that number of years again, three quarters of it will be lead, and so on. So by measuring how much uranium and how much lead are in the rock at any point in time—and assuming the laws of nature don't change in time—you can figure out how long it's been since the rock was pure uranium. About twenty years before Hubble made his measurements on galaxies, some chemists dug up a few rocks, part uranium and part lead, and used them to estimate the age of the earth. It came out to about four billion years, nearly half the age of the universe, according to Hubble. In other words, the figure that astronomers get by looking at far-off galaxies through a telescope is roughly the same as what chemists and geologists get by looking at rocks under their feet. It amazes me how those two numbers agree."

"An interesting story," said my great-grandfather. "The faith of one scientist holds up the faith of another. That's good. But it's still faith, as you were saying before. You can measure your atoms and galaxies until hell freezes over, but I doubt if you're going to know for sure how old the universe is, or even if it has an age."

"Not by being there at the start," I had to admit. "The entire recorded history of human beings goes back only ten thousand years. Our whole species goes back only a hundred thousand."

I was getting drowsy, and the fire was low. As I lazily rose from my chair to put another log on the fire, I turned and noticed that Papa Joe was also beginning to fade. He was standing in front of a bookshelf, lost in thought, and various titles slowly started to appear through his dissolving form—*Walden*, *The Double Helix*, *A Connecticut Yankee in King Arthur's Court*. I hoped he would come back again.

The Third Evening

The next day I had three lectures to give, which didn't go so well, and late meetings with students. It was seven o'clock by the time I got home. I began wondering if I'd ever see my great-grandfather again. To my delight, he appeared that evening in the usual place, chipper and perched in the wingback, scarcely after I'd got the pipe going. Apparently, he was getting the knack for his strange kind of travel.

"In my time I used mostly Prince Albert in that pipe," he said, taking a broad whiff of the smoke. "Named after the Queen's husband. Now there was a woman with good common sense. And she wasn't afraid to speak her mind either."

We chatted awhile about Queen Victoria, whom my great-grandfather was well up on. From there, Papa Joe moved on to the Great War, and how he nearly went broke when the prices of labor skyrocketed and he felt morally bound to stick to the costs in his contracts. I loved hearing his stories. After a few minutes, however, Papa Joe grew impatient and got up from his chair to look out the window.

"You haven't really kept your promise to tell me about stars," he said. I started to speak, but he continued. "I used to take your father out in the backyard at night to look at the stars. That's when we all lived in the stone house on Sixteenth Avenue South. Your grandfather was always too busy with business."

"Dad did the same with me when I was a boy," I said.

"What *are* stars, anyway?" asked Papa Joe.

"Well, to begin with, stars are pure gas, gigantic balls of gas, much larger than planets. Their gravity holds them in, the same way the earth's gravity keeps our air from flying off into space."

"All gas, you say. Confounded flimsy material for heavenly bodies, if you ask me. So if I dove into the sun, which is a star as I remember, I'd never hit solid ground, all the way to the center?"

"Right. Of course, you'd get burned into powder long before that, bones and all. Stars give off a great deal of heat as well as light."

"I have no plans to visit one of those balls of fire, beautiful as they are," said Papa Joe. "But tell me, young man, how can you be so sure that the sun isn't solid in the middle, with the gassy part only a covering, like the air around the earth? Have your solar scientists gotten up their courage and launched themselves into the sun?"

"Hardly," I said with a grin. "Scientists these days prefer to take their adventures through frightening equations. According to which, the sun requires a temperature of millions of degrees to keep itself inflated the way it does against the inward pull of its gravity. At that temperature, you can be sure any solid matter would be instantly vaporized. The sun's got to be gas all the way through."

"So now it's equations, after smudges of light through a telescope," said the old gentleman, jingling some coins in his pockets. "Please answer me this. Why don't stars burn themselves up, in such a high heat?"

"They do," I answered, "but not for a very long time. Stars outlive ordinary fires because they don't run on chemical combustion. You burn wood or gasoline or flammable gas, and you're getting energy only from the outer parts of atoms. If the sun ran on that kind of energy, called chemical energy, it would be out of fuel in a few thousand years. What you need in a star is a different kind of energy, called atomic energy. That's the energy you get from the inner parts of atoms. It's set free when two atoms fuse together to make a larger atom, which happens only under much higher heat than in chemical fires. In the sun, for example, atoms of hydrogen gas are continuously joining to make atoms of helium gas. Pound for pound of fuel, atomic energy is millions of times more powerful than chemical energy. It should keep the sun shining for billions of years."

Papa Joe nodded. "I love the way nature has various energies for each different purpose, like your great-grandmother with outfits for every occasion. Sometimes she wanted to shimmer and sometimes to blaze." He chuckled. "But back to the sun. Can your scientists predict what will happen after it's burned up its atomic fuel?"

"Yes. Near the finish, it should change brightness and color, swell up to hundreds of times its size now, and engulf the earth. Then, when it's entirely exhausted its fuel, it should collapse to a very dense sphere about the size of the earth, growing dimmer and dimmer and colder and colder. The outer planets of the solar system, the ones not boiled away earlier, will continue to orbit a dead central mass."

"It doesn't seem right," Papa Joe said, "the sun ending its career shrunken up but kept on, like an old general with a desk job." He sat for a while staring at the fire. "I just don't see how you can figure so far in advance," he said finally. "Last night it was billions of years in the past and tonight it's billions of years in the future."

"Some of the predictions come from equations," I replied.

"You talk about your equations as if they were the Ten Commandments. Where do they come from, anyway?" he asked.

"To be honest," I answered, "I wouldn't put complete trust in the equations either, if that's all I had to go on. But there's other evidence, observational evidence. Astronomers have looked at a great many stars, of all different ages and stages of development, and, from this, they believe they can piece together the life story of a single star."

Papa Joe thought for a moment. "That must be the same way those agricultural fellows figure out the way a redwood tree grows," he said. "From what I've heard, a redwood lives a lot longer than a man. But I guess if you studied a lot of them and saw some just planted and some throwing their first leaves and some getting old, you could get a pretty good idea how a single tree lives out its life."

The old gentleman got up from his chair and put three more logs on the fire. He remained standing comfortably by the fireplace, resting one arm across the mantel. "From what you've said," Papa Joe said, "I'd imagine that space should get darker and darker, as each star goes out one by one."

"It's not quite like that, Papa Joe," I replied. "New stars are continually being born, throughout the galaxy. The basic ingredient, gas, is everywhere, strewn between the stars. To make a star, the gas has to bunch up, which happens here and there because of all the activity in space. Once such a clump forms, it collapses under its own weight, causing it to heat up. Eventually the temperature is high enough that atomic fusions can get under way, and the thing becomes a star. We've actually seen newborn stars and the gas that produced them."

"Death followed by birth," said Papa Joe. "It seems like a law of nature. But with stars, I guess there are a lot of cold bodies left floating through space."

"The end isn't that gruesome for all stars," I replied. "The ones much heavier than our sun depart with a much grander flourish. They explode at the end, and, while donating their insides to space, they briefly outshine a whole galaxy. We call those stellar explosions supernovae."

"That's the way to go," said my great-grandfather. "I don't imagine that calm fellow Aristotle, who liked his universe undisturbed, would be happy with supernovae."

"He wouldn't be happy with a great many unheavenly bodies astronomers have recently found, a lot of them in our own galaxy. For example, there are pulsars and black holes, created by the collapse of stars that can't hold themselves up under their own weight. A pulsar is an extremely dense sphere with the mass of a star and a diameter of ten miles. It spins once around every second or less and spews out a stream of energy into space like a rotating searchlight. A black hole is a mass with such high gravity that not even light can escape from its surface. Large black holes, it's believed, chew up and swallow whole stars."

The old gentleman whistled. "It's a wonder our own solar system has got on so peaceful, with all of that spinning and spewing and chewing."

"Our stretch of the galaxy happens to be very quiet," I said. "The interesting goings-on are much farther out. Even with telescopes, some of these pulsars and black holes are the devil to find. Unlike stars, many of them shine mostly with X-rays, which the human eye can't see and which never get through the earth's atmosphere in the first place. Luckily, we've figured out how to launch small man-made moons, called satellites, which orbit the earth above the atmosphere. Astronomers have gotten into the act and begun loading their new instruments onto satellites. The way it works is, the instruments catch the X-rays coming in from a particular direction in outer space, convert them into electrical signals, change these into a kind of Morse code, and broadcast it all by radio to humans waiting below. On the ground, scientists take the information and try to reconstruct a picture of the object that gave off the X-rays."

"That certainly doesn't sound like what I remember of astronomy," said the old gentleman. "I knew a professional astronomer once. A big man named Thayer, who lived on Fifth Avenue. When it was time to do some observing, he'd pack up several days of sandwiches and good books for the cloudy nights, travel to the top of a mountain somewhere, and sit at the eyepiece of a telescope, making notes and drawings and simply enjoying the view firsthand. I wonder whether these X-ray fellows have fun in their work."

"Some of them do, at least the ones I know," I replied. "They hang up their graphs and their charts and their numbers sent down by satellite, and they stare at them, and pretty soon they start talking about these pulsars and black holes like they were cousins in Nebraska. Each one has got a name—there's Scorpius X-1 and there's 3U 0900-40 and there's Cygnus X-1, and so on. For each of them, the astronomers will tell you how many trillions of miles away it is, how heavy it's likely to be, how large it's likely to be, how fast it's spinning, what it would look like if the eye could see it, and dozens of other details. These things are real. Astronomers will never get anywhere near them. Astronomers will never even see them. But they're real. The instruments say they're real, so they're real."

"What's real and what's not is a swamp I'll steer clear of," said the old gentleman. "But I do like the faith of modern scientists in their gadgets. These black holes I'd like to hear more about, if you don't mind. You mentioned that light can't get away from a black hole, because of its gravity."

"Yes. That's why they're called 'black.' A black hole doesn't have a material surface like a star, but it has a boundary, and within that boundary any light emitted, even headed out of the hole, will be turned around and pulled to the center by gravity. The size of the boundary varies in proportion to the mass inside. For a black hole the mass of our sun, its boundary would be a sphere a few miles across."

"Wait just a minute" said Papa Joe. "I took you to say that we've picked up X-rays from black holes. How do X-rays get out from one of those things when light can't?"

"I'm sorry, I should have explained that," I said. The old gentleman was quicker than any of my students. "The X-rays from a black hole don't come from the black hole itself, but from hot gas rushing toward it. What we're looking at, or rather what our instruments are looking at, is a sort of cocoon of shining gas surrounding the black hole. Black holes with no gas around them are completely invisible. They're harder to find."

"But, of course, for you and your friends, invisibility is no handicap against seeing things," said Papa Joe, with a wave of his hand.

"That's truer than you think," I said, smiling. "Even if every black hole were bare and invisible, a great many scientists would still believe in them. The equations predict they exist."

"You keep dangling those damn equations," my great-grandfather said, and began growing dim.

"Come back one more night," I pleaded to his vanishing form. "For the equations. Just one more night."

"One more night," came a faint reply. After Papa Joe had gone, my study felt very empty.

The Fourth Evening

The next day I stayed home to prepare some lectures, but my heart wasn't in it. I spent the time reading a novel instead, sitting in Papa Joe's chair.

That night the old gentleman returned as he had promised, and wasted no time in getting to the topic of conversation. "Now, I'm not afraid of numbers, young man," he began. "A fellow in construction for forty years knows numbers." He paused. "But I don't understand about equations. And I especially don't understand why you put so much stock in them."

I got out a sheet of paper and wrote down:

$$C = 2\pi r.$$

"Papa Joe, this says that the circumference of a circle equals its radius times two, times pi, a special number close to 3.14."

"I remember that rule," my great-grandfather said.

"The real strength of equations is their logic," I said. "You start at one point, and an equation tells you what has to come next, according to logic. In the example here, you give the radius of any circle, and this equation says what its circumference has to be. I think the Babylonians or somebody figured the thing out first. They went out and measured the radii and circumferences of a whole bunch of circles, of all different sizes, and gradually realized that a precise mathematical law held every time. It saved them a lot of trouble when they found it. Equations in science are all like this, except usually much more involved. They start with some law about nature, and tell you what logically follows from that law, step by step. They give rules for how things ought to behave."

"Does every single item in the world have an equation for how it behaves?" asked Papa Joe.

"Most scientists would say yes—for the physical world, that is—although in some cases we haven't yet figured out what the equations are."

"So, if I understand you right, you believe that everything in nature follows rules. Whatever the thing is, you'll eventually find an equation for it, and it'll stand up and salute."

"But what's the alternative?" I asked. "To be constantly afraid that at any moment houses might float off the earth, or stars might change into wheelbarrows, or people might start talking backward? The world we're born into is strange enough as it is. We've got to believe that, at bottom, nature is at least rational. Scientists might not discover all the rules straight off, but we trust that we'll find them."

"I can see where that view might bring you some comfort," said the old gentleman.

"What's changed in this century," I continued, "is that we don't have a physical feeling for all of the rules we've been finding. The Babylonians could draw their circles and measure them with string to test their equation. Sir Isaac Newton could compare the prediction of his law of gravity with the observed motions of the planets. But many of the new rules deal with things we can't touch or see, and some of them plain violate common sense."

"I gave up common sense a few evenings ago," said the old gentleman, chuckling, "with the heavens all bursting apart and those things flying around the earth looking at invisible light. Your new equations and your new gadgets should be very happy together. But I still don't have any idea of these new rules you've been talking about."

"Let me give some examples," I said. "The ones that come to mind are from physics."

"Fine, but please hold to a slow trot, if you don't mind."

I got up to stoke the fire and pour us some tea. "In the first third of the century," I went on, "physicists discovered a new set of rules, brimful of equations, called quantum mechanics. Quantum mechanics concerns the behavior of atoms, and particles even smaller than atoms. One of the rules amounts to this: a sub-atomic particle can be at several places at the same time."

"Young man, you're galloping."

"I can't help it. The difficulty is that all our experience with the world is based on objects much larger than atoms. Golf balls and marbles are things you can pick up in your hands. They have edges. They stay where you put them. But as you go to smaller and smaller sizes, matter begins behaving differently. When you get down to atoms and smaller, your whole idea of a solid object falls apart. A particle that size, like an electron, doesn't act like a little sphere with sharp edges marking the boundary between itself and the rest of the world. An electron acts like a haze, a blur covering all the places it might be at the moment. If you throw identical marbles with identical aim at a wall, they'll all hit the wall in the same spot. But if you do the same with electrons, they'll hit it in many different spots. That's what the equations of quantum mechanics tell you. And those same equations have made very accurate predictions about many other things that have been measured and verified. So if you have faith in the theory—and physicists these days do—then you have to accept this slippery business with electrons. It goes against common sense, but there it is."

The old gentleman had got up from his chair again. "I'm beginning to get your meaning," he said. "Would you mind giving me some idea of how your physicists go about finding these equations and rules?"

"It's not much like the Babylonian method of trial and error. For many phenomena, we'd never stumble on the right rules that way. There are too many choices. Somehow, we've got to sniff out the trail." I paused a moment, and Papa Joe took a deep, lingering whiff of the aromas drifting his way from my pipe.

"Simplicity seems the best guide," I continued, "although nobody knows why. Scientists these days are constantly searching for the fewest and simplest rules possible. Two rules for a thing are better than three. A short rule is better than a long one. I know I'm being vague. Let me give an analogy. To scientists, nature is a vast game of chess. They see the board every now and then with their experiments, study what squares the pieces are on, and from this try to figure out the rules of play. At first, they might guess that every piece moves one square at a time, like a pawn. When this doesn't work, they'll try something slightly more elaborate, and so on, but never anything more complicated than the facts require. What's astonishing is that this kind of approach works remarkably well. It seems that nature loves simplicity.

"Take the case of the electron," I went on. "The precise equations for electrons were worked out by Professor Dirac fifty years ago. Now Dirac was a theorist, a pure pencil-and-paper man. I suspect he'd never been under the hood of a car in his life. But he had great faith in this idea of simplicity. So for the electron, he figured out the simplest and prettiest rules possible, consistent with the other rules he already knew. And his rules have held up for fifty years, tested by countless experiments. A more complicated theory would have been wrong. Out of his theory, by the way, came an unexpected prediction of a new kind of particle never before seen, a close cousin of the electron, called a positron. Professor Dirac wasn't looking for positrons; they just marched out of his equations for electrons. A few years after his prediction, real positrons began turning up in the lab."

"Remarkable," said Papa Joe.

"There are quite a few stories like that one. With every success, scientists have gotten more sure of themselves. In recent years, physicists have staked their reputations and millions of dollars hunting subatomic particles predicted by their theories."

My great-grandfather whistled softly. "I'd hate to be ruined by a positron that wouldn't come out of the brush," he said. "You know, I reckon it would be easier for me to follow you if I knew more math."

"Well, I'm cutting some corners, it's true," I said. "But you've been keeping up better than I would on something this new."

"What I admire most in these scientists," said Papa Joe, "is how they're willing to trust their equations against common sense. I don't believe I could follow the plans for a house that seemed upside down. That takes faith."

"I agree. You'd want to be darn sure of your architect. And you wouldn't move in right away."

We sat for a time without talking, listening to the faint bark of a dog down the street.

"Tell me about some other theories that seem contrary," said the old gentleman.

"You remember the black holes from last night?"

"Yes. They were my favorites."

"Black holes were predicted by Professor Einstein's new theory of gravity. According to the theory, if you went to live near a black hole and then came back to earth, you'd be much younger than if you'd stayed here. The gravity of the thing slows down time in its vicinity."

"Confound it," shouted the old gentleman. "I'll go along with your fuzzy atoms and particles, whatever they're good for. But time is time. A year is a year, isn't it? I must have misheard you."

"You didn't mishear me, Papa Joe, although I agree that the idea seems preposterous. You see, Professor Einstein's theories propose that the flow of time is not fixed, as it seems. Time depends on motion and on gravity. The effect is tiny unless you're moving at extremely high speeds or being pulled by a very high gravity, and that's why you don't notice it. But sensitive instruments and clocks have verified the effect. It's taken me years to get used to the idea."

"Now that I think of it, I remember a big commotion over one of Einstein's predictions being proved."

I nodded. "You're probably remembering the famous experiment during the solar eclipse of 1919. One of Einstein's theories predicted that light should be attracted by the sun, the way a planet is. The effect is very small, because light travels much faster than planets, but it's there and it's measurable. To test the prediction, you examine some stars just past the edge of the sun. According to Einstein, the starlight should be deflected by the sun on its way to earth, and the images of the stars should be slightly distorted. Some astronomers did the measurement at the first convenient eclipse, when stars could be seen near the sun, and confirmed the effect. These days, most scientists believe just about every prediction of Einstein's theories, even the ones not yet proved."

"I wonder whether Professor Einstein was bothered by this odd business with time slowing down," said Papa Joe.

"I don't think so," I replied. "From what I can tell, Einstein believed that the new ideas were logical and natural, given certain facts. He had this wonderful way of starting from scratch, without taking anything for granted. And he never expected to experience all the mysteries of nature with his body. To him, it was pleasure enough to get a mental glimpse now and then, and imagine the rest."

As I got up to stretch, the church clock in town chimed eleven. The old gentleman was back at his spot near the window, looking out at the night. I joined him there. Sirius, the brightest star in the sky, was in easy view, as well as half a dozen constellations—celestial pictures of hunters and serpents and lions and dogs, ancient visions of men and women looking for order.

"You know," said Papa Joe, "I believe your faith is contagious. These last few nights I've felt so tiny I could fit inside an atom, and so big I could step from one star to the next." He paused, staring out the window. "I proposed to your great-grandmother on a night like this." Papa Joe turned and took a long look around the room. "You take care of that pipe."

I stood for a moment beside my great-grandfather, shoulder to shoulder, and then he melted away.

SMILE

It is a Saturday in March. The man wakes up slowly, reaches over and feels the windowpane, and decides it is warm enough to skip his thermal underwear. He yawns and dresses and goes out for his morning jog. When he comes back, he showers, cooks himself a scrambled egg, and settles down on the sofa with *The Essays of E. B. White.* Around noon, he rides his bike to the bookstore. He spends a couple of hours there, just poking around the books. Then he pedals back through the little town, past his house, and to the lake.

When the woman woke up this morning, she got out of bed and went immediately to her easel, where she picked up her pastels and set to work on her painting. After an hour, she is satisfied with the light effect and quits to have breakfast. She dresses quickly and walks to a nearby store to buy shutters for her bathroom. At the store, she meets friends and has lunch with them. Afterward, she wants to be alone and drives to the lake.

Now, the man and the woman stand on the wooden

dock, gazing at the lake and the waves on the water. They haven't noticed each other.

The man turns. And so begins the sequence of events informing him of her. Light reflected from her body instantly enters the pupils of his eyes, at the rate of 10 trillion particles of light per second. Once through the pupil of each eye, the light travels through an oval-shaped lens, then through a transparent, jelly-like substance filling up the eyeball, and lands on the retina. Here it is gathered by 100 million rod and cone cells.

Cells in the path of reflected highlights receive a great deal of light; cells falling in the shadows of the reflected scene receive very little. The woman's lips, for example, are just now glistening in the sunlight, reflecting light of high intensity onto a tiny patch of cells slightly northeast of back center of the man's retina. The edges around her mouth, on the other hand, are rather dark, so that cells neighboring the northeast patch receive much less light.

Each particle of light ends its journey in the eye upon meeting a retinene molecule, consisting of 20 carbon atoms, 28 hydrogen atoms, and 1 oxygen atom. In its dormant condition, each retinene molecule is attached to a protein molecule and has a twist between the eleventh and fifteenth carbon atoms. But when light strikes it, as is now happening in about 30,000 trillion retinene molecules every second, the molecule straightens out and separates from its protein. After several intermediate steps, it wraps into a twist again, awaiting arrival of a new particle of light. Far less than a thousandth of a second has elapsed since the man saw the woman.

Triggered by the dance of the retinene molecules, the nerve cells, or neurons, respond. First in the eye and then in the brain. One neuron, for instance, has just gone into action. Protein molecules on its surface suddenly change their shape, blocking the flow of positively charged sodium atoms from the surrounding body fluid. This change in flow of electrically charged atoms produces a change in voltage that shudders through the cell. After a distance of a fraction of an inch, the electrical signal reaches the end of the neuron, altering the release of specific molecules, which migrate a distance of a hundred-thousandth of an inch until they reach the next neuron, passing along the news.

The woman, in fact, holds her hands by her sides and tilts her head at an angle of five and a half degrees. Her hair falls just to her shoulders. This information and much much more is exactingly encoded by the electrical pulses in the various neurons of the man's eyes.

In another few thousandths of a second, the electrical signals reach the ganglion neurons, which bunch together in the optic nerve at the back of the eye and carry their data to the brain. Here the impulses race to the primary visual cortex, a highly folded layer of tissue about a tenth of an inch thick and two square inches in area, containing 100 million neurons in half a dozen layers. The fourth layer receives the input first, does a preliminary analysis, and transfers the information to neurons in other layers. At every stage, each neuron may receive signals from a thousand other neurons, combine the signals—some of which cancel each other out—and dispatch the computed result to a thousand-odd other neurons.

After about thirty seconds—after several hundred trillion particles of reflected light have entered the man's eyes and been processed—the woman says hello. Immediately, molecules of air are pushed together, then apart, then together, beginning in her vocal cords and traveling in a springlike motion to the man's ears. The sound makes the trip from her to him (twenty feet) in a fiftieth of a second.

Within each of his ears, the vibrating air quickly covers the distance to the eardrum. The eardrum, an oval membrane about .3 inch in diameter and tilted fifty-five degrees from the floor of the auditory canal, itself begins trembling and transmits its motion to three tiny bones. From there, the vibrations shake the fluid in the cochlea, which spirals snail-like two and a half turns around.

Inside the cochlea the tones are deciphered. Here, a very thin membrane undulates in step with the sloshing fluid, and through this basilar membrane run tiny filaments of varying thicknesses, like strings on a harp. The woman's voice, from afar, is playing this harp. Her hello begins in the low registers and rises in pitch toward the end. In precise response, the thick filaments in the basilar membrane vibrate first, followed by the thinner ones. Finally, tens of thousands of rod-shaped bodies perched on the basilar membrane convey their particular quiverings to the auditory nerve.

News of the woman's hello, in electrical form, races along the neurons of the auditory nerve and enters the man's brain, through the thalamus, to a specialized region of the cerebral cortex for further processing.

Eventually, a large fraction of the trillion neurons in the man's brain become involved with computing the visual and auditory data just acquired. Sodium and potassium gates open and close. Electrical currents speed along neuron fibers. Molecules flow from one nerve ending to the next.

All of this is known. What is not known is why, after about a minute, the man walks over to the woman and smiles.

E.T. CALL
HARVARD

The first person on earth to receive greetings from extraterrestrial beings, if they come in the next few years, will very possibly be a fellow named Skip Schwartz. Mr. Schwartz watches over the Harvard-Smithsonian's eighty-four-foot radio telescope in Massachusetts. Nestled in the countryside some twenty-five miles northwest of Cambridge, the telescope is listening around the clock for an intelligible beacon from other solar systems. It is the most comprehensive search of its kind. The entire operation—including the huge antenna dish that slowly sweeps the sky as the earth moves, the various electronics that process the incoming signals, and the computers that digest the data—practically runs on its own. Each day after lunch, Mr. Schwartz, who lives nearby, comes into the tiny building next to the telescope, takes a look at the ten biggest signals received in the previous twenty-four hours, writes down the biggest of these in a notebook, and restarts one of the computer programs.

E.T.s calling or not, a certain amount of static is

expected in any radio. Furthermore, lifeless matter in space carelessly sends out radio signals of its own, and steps must be taken to weed these out from the deliberate variety. Some such precautions are built into the equipment. When Mr. Schwartz sees something that looks interesting, he calls up Harvard University and Paul Horowitz, a physics professor who heads the project and created most of the hardware. Depending on the intensity of the unidentified signals and of Mr. Schwartz's voice, Professor Horowitz will drive right over in his Saab and can, within several hours, eliminate to a high degree of confidence any signals of mundane origin that somehow slipped through. In the several years Horowitz's Project Sentinel has been operating, there have been only two false alarms: one caused by the sun and one caused by some unknown local interference, which gave itself away by showing up at eight different points in the sky. If and when Horowitz ever sees a signal that passes all the tests, he will quietly report its coordinates of origin to other radio observatories for confirmation. An independent check would come in the next couple of days. "If the E.T.s have done their work well," says Horowitz, "the signal will be unmistakably artificial," like the first few digits of *pi,* beeped out in binary, or the first few prime numbers. After that, perhaps one of their books.

In addition to Horowitz's project, there is currently (as of 1986) only one other full-time scientific attempt at communication with extraterrestrial intelligence (CETI), and only a handful of other, sporadic attempts. In fact, the entire subject is in its infancy. The first serious effort at CETI was Project Ozma in 1960, developed by Cornell astronomer Frank Drake at the National Radio Astronomy Observatory in West Virginia. Ozma, also a radio receiver, ran a total of two hundred hours and investigated only two nearby stars.

After Drake's pioneering work, scientists in the United States and Europe began meeting to discuss the subject and, in 1971, gathered at the first major conference on CETI, held in the Soviet Union with the official blessings of both the Soviet and the American national academies of sciences.

In the summer of 1967, the radio telescope at Cambridge University was listening for natural radio emissions from space when something very odd showed up on the charts: extremely regular pulses arriving every 1.33730115 seconds. Nothing like this had ever been seen before from space. For several months the Cambridge astronomers kept the news to themselves as they tried to think of an explanation. In February, the discovery, still unexplained, went public. Among the possible explanations was the beacon of an extraterrestrial intelligence, and the source of the pulses was dubbed, half jokingly, LGM, for Little Green Men. Astronomers settled on an answer within the year, having dismissed artificial causes but uncovering some rather exotic natural ones. The pulses were originating from a highly dense star called a neutron star, hypothesized but never before observed, which was spinning on its axis at the enormous rate of one revolution every 1.33730115 seconds and emitting radio waves by natural processes. More than a hundred of these spinning neutron stars have since been observed. They're called pulsars.

Assuming for the moment that E.T.s exist and would be amenable to getting in touch, there is one overriding principle accepted by all terrestrial scientists: it is fantastically cheaper for us, or for them, to make contact by sending radio messages than by sending ourselves. Suppose, for example, we wanted to survey the nearest star, about four and a half light years away (more than 250,000 times the distance to the sun). To travel by spaceship there and back within, say, ten years earth's time requires making the trip at 90 percent of the speed of light, in something like a thousand-ton cabin to house the needed crew and supplies. Even with a rocket propelled by the most efficient means theoretically possible—pure matter and antimatter conversion, which we have no idea how to design—we'd need 360,000 tons of fuel, half matter and half antimatter. That's enough energy to provide all the electrical power now used by the United States for several million years. On the other hand, to exchange two ten-word telegrams over the same distance, using Horowitz's modest-sized radio at both ends, would cost about $50. E.T.s searching for us, with any mind for economy, should come to similar conclusions. Terrestrial scientists, therefore, are listening to their radios rather than waiting for UFOs to land.

What channel should we tune in to? Giuseppe Cocconi and Philip Morrison, in a sparkling two-page paper published in 1959 and now taken as near gospel, theorized that one frequency stands out among all the rest. Considering all natural forms of radio interference in space, the band (range) between 1,000 and 10,000 megahertz is exceptionally free from unwanted static. Furthermore, within that band there is a special frequency, 1420 megahertz, which is the prominent radio frequency emitted naturally by hydrogen atoms. Hydrogen is special because it's the most abundant and simplest atom in the universe. The average technically able E.T., presumably far smarter than Cocconi or Morrison, should also notice these things and elect to broadcast at 1420 megahertz. It happens to be not much above the highest frequencies picked up by earthly televisions.

Unfortunately, an inspired guess as to the obvious channel is not nearly enough. That's because we and

E.T. are in motion with respect to each other, causing any radio waves to shift in frequency between emission and reception. This motional effect, known as the Doppler shift, is more familiar with sound waves. We hear a train's whistle rise in pitch (frequency) as the train approaches us and fall as it moves away. Here, the motions of sender and receiver—and corresponding frequency shifts—are simple. In CETI, the situation is vastly more complicated: our planet rotates on its axis and at the same time orbits our sun, which itself is moving through the galaxy. Light years away somewhere, E.T. is broadcasting from his or her planet, which is also spinning and orbiting its own central star, and so on. Radio waves might be broadcast from E.T.'s antenna at 1420 megahertz, but they would arrive here at frequencies chasing all over the dial. If well known, any of the astronomical motions and resulting Doppler shifts can be compensated for, but we know nothing about the individual motion of E.T. until we find him, nor do we know very accurately the motion of our own sun through the galaxy. The strategy of Project Sentinel is based upon the assumption that any E.T. interested in talking has specifically targeted our sun and is fully compensating for his various motions relative to it. Horowitz, then, simply has to compensate for the various motions of the earth about the sun, which are known to high precision. His electronics does that accounting instantly and on the spot. Horowitz is something of a whiz kid with electronics.

It would be highly desirable, of course, to do away with the unlikely assumption that E.T.s have singled out our particular star. That would be wasteful on their part, unless they knew we were here and worth a lot of attention. A much better bet for E.T.s would be to broadcast randomly into space until contact is made somewhere, and a better strategy for us would be to allow for some uncertainty in the Doppler shift of this beacon. That means we have to search many channels around 1420 megahertz. Horowitz's new Project Meta, which was unveiled at Harvard in September 1985, is doing just that. Costing $95,000, Meta has two hundred times the frequency coverage of Sentinel. Both Meta and Sentinel operate on a budget of $20,000 per year (about the cost of two Oldsmobiles) and have been funded by the Planetary Society, a private organization recently founded by planetary scientists Bruce Murray and Carl Sagan. Sometime in the next decade, NASA plans to build a much larger receiver. No one in this business seems in a hurry. Round trips for messages, even at the speed of light, may be thousands of years or more.

One of Sentinel's significant technical advances was its novel ability to focus on an ultranarrow band of frequencies around the target frequency. No radio receiver can limit itself to a single frequency, but the narrower the band, the weaker the background noise within that band—just as the smaller the bull's-eye, the fewer darts in a random throw of ten will hit it. Meta has these same capabilities, but can cover many more bands.

Sentinel could pick out a nonrandom signal of one megawatt power broadcast from a dish its size seven light years away. On galactic scales that isn't terribly far, but then eighty-four feet isn't much for a radio dish, and a megawatt is not much more power than some earthly TV stations put out. Transfer Sentinel's electronics to the thousand-foot dish at Arecibo, Puerto Rico, and it could hear a megawatt broadcast from its twin at one thousand light-years. Within that range shine a million stars similar to our sun. Around some of those stars might be planets like earth, planets with life.

In October of 1984, Horowitz invited me to visit
Project Sentinel. When I arrived with him at the site
of the telescope, I was greeted by an enthusiastic engi-
neer named Mike, who was perched forty feet up in
a crane, scraping rust from the telescope pedestal.
Mike, dressed in blue jeans and tennis shoes and
plugged into a Sony Walkman, had come from Seattle
to donate a few weeks of his time to Project Sentinel.
The sixty-foot pedestal badly needed painting, and
Professor Horowitz soon became engaged in discussing paint colors with another short-term volunteer, an
inventor from Martha's Vineyard named Mal. "So it's
going okay?" Horowitz asked Mal. "Yeah. Sandpaper
works good on the pipes."

I looked around the little operating station next to
the telescope. On one of the walls inside hangs a tapestry made for Sentinel by third-graders at the nearby
elementary school. The tapestry is richly decorated
with pictures of extraterrestrials of all imaginable
shapes and colors and numbers of eyes and legs, and
it says in large letters, "E.T. call Harvard." While I
was standing in the data room, a large signal came in
but didn't hang around for more than one thirty-
second interval of data collection. "That's only noise,"
said Horowitz quickly. Any intentional beacon should
stay in view of the dish for several minutes. I must
admit I was disappointed, for the third-graders and
myself. Even a good false alarm would have been welcome. Horowitz's operation, with its various safe-
guards and its technical prowess, runs too smoothly for
drama.

It was Copernicus who demolished the view of an
earth-centered universe. In 1686, less than 150 years
later, the literary science writer Bernard Le Bovier de
Fontenelle took the new world view to its natural
conclusion in his fictionalized conversations with a

countess as they strolled through the park: "Our Sun enlightens the Planets; why may not every fix'd Star have Planets to which they give light? . . . the Earth swarms with Inhabitants. Why then should Nature, which is fruitful to an excess here, be so very barren in the rest of the Planets?" In the second decade of this century, we learned that our sun resides in the suburbs of a galaxy of billions of suns. In the 1950s, we learned that amino acids, the building blocks of proteins and terrestrial life, form naturally under the probable conditions of the primitive earth. In 1984, astronomers observed what appears to be a planetary system forming around a nearby star, a process thought to be common. It's hard to believe no one else is out there. Ours is the first generation of earthlings capable of finding out.

Horowitz insisted that before leaving I climb out on the dish. To get there, we had to first ascend the stinger tube, an inclined enclosure angling sixty feet up to the bottom of the dish. The dish itself was aimed straight up, an eighty-four-foot-wide bowl of aluminum mesh that noticeably sagged underfoot. Against my better judgment, Horowitz persuaded me to crawl upward along that steeply curved bowl the final fifteen feet in the air and take a look over the rim. Stretching around me was the New England countryside in full fall foliage, with little farmhouses scattered here and there, and, far off in the distance, the faint image of Boston. And above that, a blue, blue sky.

RENDEZVOUS

Beginning with the blast-off of two Soviet rockets, last month, a total of five spacecraft are being launched from the planet Earth to rendezvous with Halley's Comet. The comet makes its once-in-a-lifetime visit to the solar neighborhood about every seventy-six years, and, this trip, it will be nearest the sun on February 9, 1986. The spacecraft will come up with it about a month later. Until now, astronomers have had to settle for earthbound views of Halley's and the thousand-odd other comets recorded over the centuries.

Comets, throughout history, have stirred up all sorts of speculations and fancies: views of comets as pieces of planets, as orbiting hailstorms, as dirty snowballs; views of comets as omens of defeat in battle or of the untimely demise of an emperor; views of comets as carriers of flu, as sowers of life on earth; theories that it was a comet that signalled the Norman invasion, that served as the Star of Bethlehem, that brought on the bubonic plague, that blackened the sky and killed off the dinosaurs; comparisons of a comet's head to the eye of an ox and its tail to the fan of a peacock; drawings of angry comets splitting the earth while the moon looks on merrily; sweepstakes on guessing the date of a comet's arrival. Fitted out with various scientific instruments, the five spacecraft—the two from Russia (both of which contain United States contributions), two from Japan, and one from a consortium of countries in Europe—are expected to find out at close range exactly what a comet is and does, and to radio their data back to Earth, ninety-one million miles away.

Halley's Comet is special in several respects. Its orbit, which reaches past Neptune at the farthest point, is still sufficiently small to have been well charted, but its flashy display is more like that of "younger" comets, of much larger orbits and less frequent appearances. Halley's is the first comet whose periodic sightings were reckoned to be returns of the same object. Taking the credit for this idea was Edmond Halley, scientist extraordinaire and general snooper: an Englishman who measured the areas of counties by cutting up the map and weighing the pieces, and who figured out a scheme for the first life-insurance tables; a man who persuaded his colleague Isaac Newton to finish the "Principia" and dogged him into inserting an exhausting section on comets in its second edition. For thousands of years before Halley's work, people believed that every comet streaking through the solar system was on a one-way trip. When Halley took up comets, his friend Newton, having just discovered the law of gravity, explained to him that comets, like all bodies orbiting the sun, ought to have trajectories in one of four shapes: hyperbola, parabola, circle, or ellipse. The first two don't turn around and close on themselves; the last two do. Halley looked into the latter possibility and convinced himself that comets recorded in 1531,

1607, and 1682 (this one seen by him at age twenty-six) were the same object, orbiting the sun in an elongated ellipse. Halley's triumph, which he was not around to enjoy, was the dutiful return of his comet in December of 1758, as he had predicted.

Last time through, in 1910, Halley was three days behind schedule. This couldn't be explained just on the basis of a planetlike orbit about the sun, and gave rise to the prevailing "dirty snowball" theory of comets. According to this theory, deep inside a comet's head is a solid nucleus, a few miles in diameter, that is made of ice, dust, and minerals. As a comet nears the sun, the ice and dust on its sunny side evaporate and stream off, nudging the comet in the opposite direction. This additional rocketlike force on a comet can account for its off-cue behavior. These same vapors boiled out of the tiny nucleus create the comet's head and its tail, which can trail on in a flimsy wisp for many millions of miles. Zipping through the solar system as it does, with a body larger than the sun's and a total mass far smaller than the moon's, a comet has more style than substance. No one knows for sure how big the central snowball is, or even whether there is one, but the cameras on the spacecraft should find out.

The thing that many scientists will be waiting for is news of precisely what's inside a comet's nucleus. If comets condensed out of the same embryonic gas and dust as our solar system did, they will still contain those ingredients in their original form. Planets and moons, heated by the sun and squeezed by their own weight, have melted and congealed, and so blurred the record of their birth, but comets spend most of their time far from the sun and have little weight to support. Comets, therefore, could shed light on the origin of the solar system.

As I, like others, look forward to the beamed-down

photographs and chemical readouts of Halley's Comet, what strikes me is the mission's lack of practical value. At a cost of hundreds of millions of dollars, the world will not learn new methods for producing plastics, will not get back better silicon computer chips grown in space or new communication networks. The countries involved are not likely to boost their national prestige nearly so much as we and the Russians did in the frantic post-Sputnik race to space. And comets are poor places to put military bases. All we'll get is a better hold on this strange universe we find ourselves in—like my four-year-old daughter, who knows nothing of plastics or politics but who asks questions about the world ("Why is the sky up and not down?" "Where does the rain come from?") and is comforted and delighted by the answers.

TIME FOR
THE STARS

More than once in the past year, I've been drawn into a heated discussion over the vast sums being budgeted toward a military defense in space. Much of this money has been suddenly offered to the scientific community. Should it be accepted? For me, in addition to the ethical and practical questions, this raises the issue of applied science versus pure science.

I am worried that our country has become increasingly shortsighted to the value of pure science. One recent example was the court-ordered breakup of American Telegraph and Telephone, leaving its basic research group, Bell Laboratories, in a vulnerable spot. Another was the congressional veto of a relatively cheap exploratory mission to Halley's Comet, which visits the solar system only once in a lifetime. Certainly, few people would deny the material comforts, the economic advantages, the power to make war or peace that applied science brings. But in pursuing these other goals, we have paid less and less attention to the value of science for its own sake.

Our national preoccupation with applications goes

back to the cultural and political origins of our country. In Europe, science was traditionally considered a part of culture, and a person could devote his life to science as a gentleman and a scholar. Isaac Newton, as a fellow of Cambridge University, needed no justification for his studies of physics. Carl Friedrich Gauss, who made brilliant contributions to mathematics and astronomy in the early nineteenth century, was supported through the patronage of the Duke of Brunswick. In contrast, when science got under way in America, in the middle 1800s, the democratic ideals of our young country demanded a direct accounting to the people, a direct benefit to society. Scientific research was usually supported only if it was part of a practical or technical enterprise, like the U.S. Weather Service, founded in 1870, the U.S. Geological Survey, founded in 1879, or the National Bureau of Standards, founded in 1901. Gradually, our nation began to take pride in and identify with its technological achievements (which exclude pure science by definition). The American hero of science is Thomas Edison, not Willard Gibbs, who made fundamental contributions to the theory of heat. During World War I, even the great physicist Robert Millikan said that "if the science men of the country are going to be of any use to her, it is now or never." Since World War II, of course, our country and all countries have been keenly aware that military might can be gotten through science.

Also since World War II, science has become a big business. In many areas of science, the romantic days of the lone scientist, uncovering the secrets of nature with homemade equipment, are gone. Experiments today often require large teams of scientists, large budgets, and large bureaucracies to manage them. Some of these operations could not be mounted unless they utilized the existing military-industrial complex. And the pace of society in general has quickened. Under constant pressures, we grasp for the short-term payoff.

Why should our nation, or any nation, support pure science? Why should a nation pay for an activity that brings it no clear economic or military advantage? Why should a nation support an activity that seems *useless*?

It seems to me that pure science has several different values. In order of increasing range into the future, pure science entertains us, it provides the soil from which technology grows, it changes our world view, and it grants us cultural immortality.

On an immediate, day-by-day basis, learning new things pleases us, and there is no doubt that we learn from pure science. Furthermore, what we learn is true, it concerns the real world, and it can be understood in broad terms by every intelligent person. Nonscientists are entertained by learning what comets are made of, just as they are entertained by seeing a new Neil Simon play or reading a new book by Gabriel García Márquez. Everyone is a potential consumer of pure science. If pure science cannot pay for itself in the marketplace, as movies and books do, it is perhaps because its pleasures lie in knowledge. Still, this knowledge brings a special kind of happiness, and the happiness of a nation's people counts for something.

Pure science may seem useless in the usual sense, but over a long period of time it surely leads to economic and technological benefits. If we stop paying for pure science today, there will be no applied science tomorrow. Darwin's work on evolution and Mendel's on the heredity of plants laid the foundations for the science of genetics, which eventually led to the discovery of DNA, which led to genetic engineering, which is now exploding with unimaginable applications. Faraday's discovery of how a magnet can produce electricity made possible the first hydroelectric power plant, fifty years later. Yet Darwin and Mendel and Faraday were not supported with any such profits in mind, nor could they have been. A nation cannot bet on pure scientists like betting on horses. It can, however, build stables. I remember a Robert Heinlein novel about a research outfit called The Long Range Foundation. The Long Range Foundation was chartered as a non-profit corporation, dedicated to future generations. Its coat of arms read "Bread Upon the Waters," and it prided itself in funding only scientific projects whose prospective results lay at least two centuries away. It was happy to waste money. Unfortunately, the directors could never do their job right, and the foundation's most preposterous projects quickly began piling up embarrassingly large profits.

The third value I mentioned is the ability to change our world view. This quality is often subtle, but its importance cannot be overestimated. I think Henry Adams understood the value of pure science when he wrote, in the early 1900s, that Madame Curie's discovery of radioactivity suddenly made the unknowable known. Since ancient times, Western man had worshiped this ultimate material unit called the atom —indestructible, impenetrable, exquisitely unfathomable. Then, at the end of the last century, Madame Curie found that atoms of radium hurled out tiny pieces of themselves, and our view of nature would never be the same again.

It might be helpful to give a couple of examples of this in more detail. I will choose from astronomy, which is the most useless science I know and my personal profession. Actually, astronomy was once highly practical. Early civilizations used it for tracking the seasons, planting crops, and navigation. Since then, astronomy has advanced to its present condition.

As a first example, consider Kepler's discovery that the orbits of the planets are elliptical. Before Kepler, there was universal agreement, dating back many centuries, that the orbits of heavenly bodies were circular. To defer to Aristotle, whose opinions on many things molded the Western world view, the circle was the natural figure for heavenly motions because of its uniqueness and perfection. Only circular orbits were proper for the divine and eternal planets. In fact, Aristotle arranged the entire cosmos in a sequence of rotating spheres, centered on the earth. Once nominated, the circle showed great staying power. When people later noticed that the planets changed in brightness—and hence distance from earth—during their orbits, astronomers invented an elaborate set of circles upon circles, whereby each planet performed a small circular orbit about an imaginary point that itself traveled in a large circular orbit about the earth. Even Copernicus, who demolished the idea of an earth-centered cosmos, clung to the idea of circular orbits.

Kepler had the good fortune of being the student of Tycho Brahe, a wealthy Danish astronomer who spent night after night observing the planets from his private island. Brahe's naked-eye reckonings of planetary positions were the most accurate ever taken. Kepler inherited this gold mine of data around 1600. His job was to make sense of it. In addition to having good material to work with, Kepler owed his success to two other factors: he was a dedicated Copernican, and he believed in the Platonist ideal that nature follows mathematically simple laws. What were the laws governing the motions of the planets? What were the shapes of the orbits? Kepler struggled with countless trial orbits of compounded circles. Eventually, he was forced to admit that they just wouldn't fit Brahe's data. Then he discovered ellipses. (Every artist knows the ellipse; it is a foreshortened circle.) One ellipse for each planetary orbit was also much simpler than two circles. The sacred circle had been replaced by the accurate and economical ellipse.

Kepler's success gave strong support to the Copernican system, in which the earth is simply another planet, orbiting the sun. We know that Newton, as a student, studied Kepler. When Newton presented his incomparable *Principia* to the Royal Society in London, it was introduced as a mathematical demonstration of the Copernican hypothesis as proposed by Kepler. Newton's *Principia* in turn, with its laws of motion and gravity and its unflagging application of these laws to pendulums and planets, provided a firm scientific foundation for Descartes' view of the universe as a giant mechanical clock. After Kepler and Galileo and Newton, nature became rational.

My second example of how pure science changes our world view is the fairly recent discovery that the universe is expanding. The galaxies are flying away from each other. When this observed motion is mentally played backward in time, the galaxies crowd closer and closer, stars and planets and even atoms are squeezed together and disrupted, until, some 10 billion years ago by the best estimates, the entire contents of the now-visible universe were compressed to a size smaller than an atom. That was the beginning of the universe. It is called the Big Bang.

Virtually every culture in recorded history has had its myths about the origin of the universe and when that origin occurred. Many have believed in no origin at all. Aristotle, for instance, gave numerous philosophical arguments why the universe had to be unchanging and everlasting. One of his arguments went something like this: If the universe had a beginning at some finite time in the past, then there would have been an infinite time before that when the universe did not exist, but had the potential for existing. However, the nonexistent universe could not have slumbered for an infinite time in such a state of pure potentiality. Therefore, the universe has always existed in its present state of perfect composure. Isaac Newton arrived at a similar conclusion by a somewhat more scientific (but still erroneous) approach. Newton argued that if the universe were expanding or contracting, there would have to be a center about which such motion took place. In an infinite space, however, no position in the universe should be so privileged. Therefore, the universe must be always at rest.

The discovery of what the universe is actually doing came in the 1920s. Using a large telescope and various instruments, the astronomer Edwin Hubble was able to determine that other galaxies are moving away from our galaxy, with speeds proportional to their distances. The closer galaxies are retreating from us more slowly than the farther ones. This is exactly the situation for dots painted on the surface of an expanding balloon. From the vantage of each dot, representing a galaxy, it appears that the other dots are moving radially away from it, at speeds proportional to their distances. The view is the same from any dot, and no dot is the center. That was Newton's mistake. Newton didn't realize that expansion could occur about *every* point in space. He didn't have the right picture in his mind. He also didn't have much equipment. I think that if Newton were here now, or Aristotle, or Moses Maimonides, or Francis Bacon, they would sit still for a lecture on the origin and motion of the universe.

It is too early to know the consequences of our

discovery that the universe is expanding. There is no doubt, however, that our world view has been changed. One sign is that Einstein insisted at first on a static universe—even when his own cosmological equations naturally predicted a universe in motion. For centuries before Hubble, the majestic tranquility of the heavens symbolized the eternal and the immutable. That soothing symbol is now gone.

One can list many other discoveries that are too new to judge. What are the consequences of learning that time flows at a variable rate, depending on the motion of the clock, or that all life forms on earth get their blueprints from the same four molecules? I don't know, but I am certain that these recent discoveries have begun to seep through our culture and alter our thinking.

Discoveries in pure science are not just about nature. They are about people as well. After Copernicus, we have taken a more humble view of our place in the cosmos. After Darwin, we have recognized new relatives hanging from the family tree. We need to be periodically shaken up. We need periodically to break free from the endless cycle of one generation passing dimly into the next, one human lifetime after another. We got stuck some centuries back, and it was called the Dark Ages. Changing our world view helps us break free.

I come now to cultural immortality, which, of course, transcends individual nations. To quote Thoreau:

In accumulating property for ourselves or our posterity, in founding a family or a state, or acquiring fame even, we are mortal; but in dealing with truth we are immortal, and need fear no change or accident.

Pure science deals with truth, and there is no greater gift we can pass to our descendants. Truth never goes out of style. Hundreds of years from now, when automobiles bore us, we will still treasure the discoveries of Kepler and Einstein, along with the plays of Shakespeare and the symphonies of Beethoven. The civilization of ancient Greece has vanished, but not the Pythagorean theorem.

Several years ago, I went to Font-de-Gaume, a prehistoric cave in France. The walls inside are adorned with Cro-Magnon paintings done fifteen thousand years ago, graceful drawings of horses and bison and reindeer. One particular painting I remember vividly. Two reindeer face each other, antlers touching. The two figures are perfect, and a single, loose flowing line joins them both, blending them into one. The light was dim, and the colors had faded some, but I was spellbound. If our civilization can leave something like that for posterity, it will be worth every penny.

FOUR FINGERS
IN A HUNDRED
CUBITS

In November of 1907, Einstein was sitting in a chair in the patent office in Bern, Switzerland, when he had a religious experience. As he later described it:

> . . . all of a sudden a thought occurred to me. "If a person falls freely [like after jumping from a roof], he will not feel his own weight." I was startled. This simple thought . . . impelled me toward a theory of gravitation.

Eight years later, Einstein published his new theory of gravity, called general relativity.

How did Einstein get from his simple thought about free fall to his theory of gravity? The observed facts about weightlessness simply aren't enough. I tried to prove they were, for my doctoral thesis, and failed.

Einstein's leap of faith in 1907 was embodied in something he called the equivalence principle. Roughly speaking, this principle conjectures that gravity is physically equivalent to acceleration. The equivalence principle is one of those few precious insights that light up each science, like Darwin's "survival of the fittest" or Dalton's "atom." If you want to understand evolution, you need to know about survival of the fittest, and if you want to understand gravity, you need to know about the equivalence principle. The equivalence principle not only led to Einstein's theory of gravity; it also banished forever the ancient belief in absolute motion. But I'm getting ahead of myself.

In November of 1907, in the last days of his job as a patent clerk, Einstein was mulling over a very familiar aspect of gravity: all objects fall equally under the influence of gravity. Drop any two objects from the same height, and, if air resistance is small, they will hit the ground at the same time. This curious fact about gravity had been known a long time. Isaac Newton built it into his law of gravitation in the late seventeenth century. Earlier, in 1592, Galileo wrote in his *De Motu:*

The variation of speed in air between balls of gold, lead, copper, porphyr, and other heavy materials is so slight that in a fall of 100 cubits [about forty-six meters] a ball of gold would surely not outstrip one of copper by as much as four fingers. Having observed this, I came to the conclusion that in a medium totally void of resistance all bodies would fall with the same speed.

Later experiments, eliminating air resistance entirely, were more precise. In the late nineteenth century, Lóránt Eötvös, a Hungarian baron, announced that his studies with plumb bobs showed that the acceleration of gravity on different objects could not differ by more than a few parts in a billion.

This simple fact about gravity, which was taken for granted by Newton, was considered by Einstein profound. He reasoned in the following way: consider an observer at rest in a uniform gravity, pulling down. This observer drops some balls and sees them move downward, all with the same acceleration, say 32 feet per second per second. (I remind the reader that acceleration is the rate of change of velocity, just as velocity is the rate of change of distance. A car moving at a velocity of 32 feet per second changes its distance by 32 feet every second. A car *accelerating* at 32 feet per second per second changes its velocity by 32 feet per second every second.) Now consider another observer, *in a space free from gravity,* accelerating upward at 32 feet per second per second. When this observer drops some balls, she also sees them move downward, relative to her, all with the same acceleration of 32 feet per second per second. The second observer sees all dropped objects accelerate equally because she herself is accelerating. The first observer sees all dropped objects accelerate equally because of the nature of gravity. Yet the observations in each case are the same. Neither person can tell, from the motion of dropped objects, which of the two situations she is in. Everyone has experienced the similarity between gravity and acceleration. When you accelerate forward in your car, you are pushed back in your seat, just as if an extra, sideways gravity were pulling you in that direction.

At this point, Einstein remarks (and I quote from his later paper "On the Influence of Gravitation on the Propagation of Light," published in the *Annalen der Physik* of June 1911):

This experience, of the equal falling of all bodies in the gravitational field, is one of the most universal which the observation of nature has yielded; but in spite of that the law has not found any place in the foundations of our edifice of the physical universe.

But we arrive at a very satisfactory interpretation of this law of experience, if we assume that the systems K and K' [the situations of the first and second observers] are physically exactly equivalent.

By "physically exactly equivalent," Einstein means far more than equivalence regarding the motion of balls and the laws of mechanics. Newton could have stated that kind of equivalence. What Einstein has postulated is equivalence with regard to *all* physical processes in nature, such as heat, electricity and magnetism, nuclear phenomena, and so on. This was true inspiration. In 1907, and still today, all physical processes and laws of nature are not even known, much less tested with and without gravity. Yet Einstein has postulated that no possible experiment, no conceivable physical process, can distinguish uniform gravity from a uniform acceleration in the opposite direction. This is the equivalence principle. Einstein later called it "the happiest thought of my life."

With the equivalence principle alone, and no mathematics, it is possible to deduce new gravitational phenomena. For example, consider the trajectory of a light ray passing the sun. Objects with mass, such as planets, are attracted by the sun, but what about things without mass, such as light? Imagine a light ray emitted by a distant star and traveling by the sun on its journey to earth. Next imagine that near the sun, falling freely through space, is an observer. Our observer will naturally be accelerated toward the sun by the pull of its gravity. While this is happening, she does experiments. Now use the equivalence principle. If our observer is falling toward the sun with an acceleration of so many feet per second per second, this acceleration is equivalent to an equal gravitational pull in the *opposite* direction. Thus, our observer has the sun's gravity pulling her one way and the equivalent gravity pulling her the opposite way. The two pulls exactly cancel. She feels weightless. Note that our observer hasn't come to a sudden stop; she is still falling toward the sun. But all gravitational pulls on her have disappeared. If she drops a shoe, it will float by her side. This is the state of affairs for anyone falling freely.

Now suppose the light ray passes our observer. What does she see? There could be only one external force on the light ray, namely gravity. But for her, gravity has vanished. So she must see the light ray travel in a straight line. Light rays with no forces acting on them travel in straight lines.

Finally, analyze the situation from the earth. What is the path of the light ray from our point of view? It must bend toward the sun. Otherwise, the light ray could not travel in a straight line relative to the space-borne observer, who is falling toward the sun. The conclusion is that a light ray passing by the sun, or any gravitating body, is deflected toward that body, just as if light had mass.

This effect has been seen. It was first measured in the solar eclipse of 1919, by a team of scientists under the direction of the British astrophysicist Arthur Eddington. An eclipse is needed for the measurement, or else starlight passing near the sun won't be visible. The idea of Eddington's experiment, which was first proposed by Einstein himself, was to photograph a group of stars near the sun during the eclipse and then to compare that picture to a photograph of the same stars taken when the sun is nowhere near (like at night). In terms of its effect on the images of stars, the sun acts like a glass lens. When the sun is present, the images of stars will shift relative to each other, just as passing a lens over newspaper print will distort the print near the edge of the lens.

Eddington's eclipse expedition returned with data supporting Einstein's prediction, in quantitative detail. On November 10, 1919, on page 17 of *The New York Times,* the following headlines appeared: "LIGHTS ALL ASKEW IN THE HEAVENS, Men of Science More or Less Agog Over Results of Eclipse Observations, EINSTEIN THEORY TRIUMPHS."

As another application of the equivalence principle, consider the color of light in the presence of gravity. Assume that yellow light is emitted on the first floor of a house and is received on the second floor; that is, the light travels upward against gravity. What will be the color of the received light? Yellow, of course, if gravity had no effect. To discover that effect, apply the equivalence principle. Replace the downward-directed gravity with an upward-directed acceleration. The house, now detached from earth, is accelerating upward through space. Remember now that an accelerating object is constantly increasing its velocity. Therefore, between the moment that the light is emitted on the first floor and the moment it is received on the second, the second-floor observer has picked up speed away from the source of the light. The effect, in this case, is the well-known Doppler shift. The color of the light must shift to a lower frequency, just as the pitch of a train's whistle drops to a lower frequency when the listener is moving away from the train (or the train is moving away from the listener). Red light has a lower frequency than yellow, so the second-floor observer will receive light that is shifted in color toward the red end of the spectrum. Returning to the original situation, we have concluded that light traveling upward in the presence of gravity will be shifted in color toward the red. This is sometimes called the gravitational redshift.

The first definitive confirmation of the predicted

gravitational redshift was reported in 1954, a year before Einstein died. Daniel Popper, an American astronomer, measured the change in colors of light emitted by the white dwarf star 40 Eridani B. White dwarf stars are extremely compact stars, with a gravity much higher than the earth's or the sun's, and thus provided the first possibility for observing the effect.

Just as with "the survival of the fittest," one could go on with more predictions based on the equivalence principle. However, the deepest consequence of this principle lies not in specific gravitational phenomena, but in the relativity of motion and the final destruction of absolute space. That is the problem that Einstein next turned to, in the years between 1907 and 1915.

Until Einstein, the notion of absolute space and absolute motion was ingrained in science and culture. Newton believed in an absolute space. So did Aristotle. So did the Babylonians. What does this concept mean? Absolute motion is motion that can be determined and measured *without reference* to anything outside of the object in motion. Absolute space is the space surrounding an object in absolute motion. An example will make this easier to understand. Suppose you are sitting in a train at the station and another train rolls in. If you look at the other train, and nothing else, it sometimes seems that you are moving and the other train is standing still. To convince yourself that you are the one at rest, you can look out the window at the station. In other words, your determination that you are at rest is based upon your motion *relative* to the station. Now, imagine that your train is transported to space. You can no longer look out the window at the station. Imagine further that no stars or planets are visible or detectable in any way. How can you measure your motion now? According to the concept of absolute space, empty space by itself offers some means of deciding whether you are at rest, or moving with a velocity of so many feet per second or an acceleration of so many feet per second per second. According to the concept of absolute space, empty space contains some preordained frame of reference against which all motions can be measured.

Aristotle's belief in absolute space appeared in his doctrine that each type of substance had a fixed and natural place in the universe. In *On the Heavens* he writes: "A body moves naturally to that place where it rests without constraint, and rests without constraint in that place to which it naturally moves." The place of earth is at "the [absolute] center of the universe." This God-given center then provides a point of absolute rest, against which all other motions in the universe can be measured.

Newton had to invent an absolute space in relation to which accelerations are measured. Otherwise, the accelerations that appear in his equations of motion are not defined. In "The Scholium" of *The Principia*, Newton clearly states his views on this subject:

Absolute space, in its own nature, without relation to anything external, remains always similar and immovable. . . . We may distinguish rest and motion, absolute and relative, one from the other, by their properties, causes, and effects.

This ethereal concept of absolute space and absolute motion, without relation to anything material, bothered Einstein. Despite the vast philosophical implications of his work, despite the elegance of his theories, Einstein was actually a pragmatist. He based his thinking on tangible balls and clocks and rulers, even if they were often inside his head. As Einstein writes at the beginning of his paper "The Foundation of the General Theory of Relativity," published in *Annalen der Physik* in 1916:

> The law of causality has not the significance of a statement as to the world of experience, except when *observable facts* ultimately appear as causes and effects. . . . [Therefore], of all imaginable spaces in any kind of motion relatively to one another, there is none which we may look upon as privileged *a priori*.

That is, forget about finding a privileged state of rest or of motion, independent of physical circumstances. Stop looking for the imaginary train station in space. Look for the other train. All that matters, all that can be measured, is *relative motion* between objects of substance.

In 1905, before the equivalence principle, Einstein took a first step toward eliminating the idea of absolute space, with his special theory of relativity. A basic principle of special relativity is the physical equivalence of all observers moving at constant velocity with respect to each other. All such observers experience identical laws of nature. There is no physical way for one observer to determine, for example, that he is at absolute rest while the others are not. The theory, however, is restricted to those observers moving at *constant* velocity relative to each other, and fails to apply for accelerated observers. That is the origin of the word "special."

But what about accelerated observers? Do they experience different laws of nature? If so, then they could announce, without looking outside the window, that their motion is accelerated. Acceleration could be determined in absolute terms, without regard to any external object of substance.

Einstein was deeply troubled by the failure of special relativity to include accelerated motions. Absolute motion, of any kind, didn't make sense to him. He needed a good reason for ruling it out. And he found one, in the equivalence principle. The equivalence principle states that it is physically impossible to distinguish between a uniformly accelerated observer in a space free of gravity and a static observer in a uniform gravity. There is no way the first observer can decide that he is accelerated rather than sitting at rest in the presence of gravity. As Einstein writes in 1911—and his pleasure can be read between the lines: "This assumption of exact physical equivalence makes it impossible for us to speak of the absolute acceleration of the system of reference, just as the usual [special] theory of relativity forbids us to talk of the absolute velocity of a system." Absolute acceleration, and with it absolute space, were no more.

In 1915, after many false starts, but guided by the equivalence principle, Einstein produced his general theory of relativity. General relativity applies to all kinds of motion and shows how the laws of nature are identical in all. No particular motion is privileged above any other. General relativity is a theory of gravity, as well as a theory of motion. The equivalence principle unites the two.

I said earlier that Einstein was a pragmatist. Yet he was also an inventor of ideas. In some ways he dared more than Newton, whose theory of gravity and absolute space he replaced. Newton's famous credo "Hypothesis non fingo" (I frame no hypotheses) expressed the empirical approach to science, the cautious reliance on data that were so needed for giving birth to the modern scientific method. But after two centuries, science had grown up and gained confidence. In the equivalence principle, Einstein framed an hypothesis. It has not been disproved.

A DAY IN
DECEMBER

Shortly before six o'clock on Thursday morning, December 6, 1979, someone's dog, let loose, ran yelping down Embarcadero Road in Palo Alto, turned right on Waverley, and dropped from fatigue and boredom near the intersection of Santa Rita Avenue, having woken all sleepers within earshot. It was still dark. Lights blinked on one by one along the animal's path, people groped for robes and went to the toilet, and another day began.

By half past seven University Avenue was filling with college students cycling to their early classes. At the doorway of one of the large homes on Waverley near University, a woman in her early forties, wearing a smart tweed suit, called out to her husband, "George, don't forget the gardening book for Betty." George, in pinstripes, nodded and drove away to a Silicon Valley company.

A couple of hours later, in his rented house on Camino a Los Cerros, Alan Guth got out of bed, had two hard-boiled eggs, and waved good-bye to his wife and son (who the day before had said, "Daddy's home," for the first time). On his ten-speed, which was kept in good repair with supplies from the Palo Alto Bike Shop, he rode southeast to Sharon Road, turned right, flew past the shopping center, turned left onto Sharon Park Drive, turned right on Sand Hill Road, and entered the grounds of the Stanford Linear Accelerator Center. His office was on the northeast corner of the third floor of Central Laboratory, in the theory group. Guth was a thirty-two-year-old physicist.

By this time it was half past ten. Students and student-types were beginning to hang around Printer's Inc., a bookstore on California Avenue with a coffee bar and classical music in the background. A well-fleshed man in corduroys was thumbing through *Diet for a Small Planet,* wondering what to serve at his vegetarian dinner party.

On the street outside, the day was fine, unseasonably fine. The woman in the smart tweed suit, on her way to look at new wallpaper, decided to go home and change into something cooler. The weatherman had predicted rain. She hurried. Her old wallpaper of seven years, brimming with five-inch burgundy squares caught within a thicket of yellow diagonal stripes, had to go.

Guth started work with coffee. His colleagues on the third floor shared a community coffee pot for $3 a month per person. Around noon, after placing an anxious phone call about a possible job for the coming year, Guth went with two friends to lunch at the New Leaf. Afterward, back in his office, he wrote some correspondence—he did all his writing with a Radiograph pen, with its bold, neat lines—and later discussed magnetic monopoles and cosmology with a colleague. At six o'clock Guth pedaled home. Cedar, Camino de Los Robles, Monterey, Manzanita, Camino a Los Cerros. He knew the side streets on his route.

In fifteen minutes he was home, had a broiled steak, medium rare, and after dinner he and his wife did laundry. He was out of underwear.

The man in corduroys with the dinner party outdid himself. At evening's end, he slumped exhausted on his sofa, leaving the dirty plates and enameled soufflé pans in stacks around the kitchen. Half an hour of television before bed would smooth him out. Click. A floor wax commercial.

Outside, above Palo Alto, the deep blue sky was blackening. Higher still, uncounted stars cut silently into the night. Sometime between eleven and twelve o'clock, sitting at his study desk with only pen and paper, Guth discovered mathematical evidence that, contrary to previous theories, the infant universe 10 billion years ago underwent a fantastically rapid expansion, just after which the matter that was to form atoms and galaxies and people came into being.

IN HIS
IMAGE

I recently came across a collection of scientific studies on the search for extraterrestrial intelligence, published a few years ago by the National Aeronautics and Space Administration. The book's foreword was written by Theodore M. Hesburgh, who is president of the University of Notre Dame and a Catholic theologian. Hesburgh recalls being asked (by a surprised lawyer) how a religious person such as himself could legitimately accept the possibility of other inhabited worlds out there in space. He answered: "It is precisely because I believe theologically that there is a being called God, and that He is infinite in intelligence, freedom, and power, that I cannot take it upon myself to limit what He might have done." Writing on the same proposition seven hundred years earlier, and speaking for many of the intellectuals of his day, the great theologian and philosopher St. Thomas Aquinas took exactly the opposite position: "This world is called one by the unity of order . . . all things should belong to one world." For Aquinas, who spent his life trying to reconcile faith with reason, God's omnipotence and goodness were better illustrated by a single, perfect world than by many, necessarily imperfect worlds.

The Reverend Hesburgh—and the many of us who share his ease with the possibility of other worlds—has not come to his point of view by any certain scientific evidence. Despite a great deal of searching, no life of any kind has turned up on Mars; no extraterrestrial communications have yet been detected from outer space; no planets have yet been found outside our solar system. What happened between Aquinas' day and ours was a revolution in how we think of ourselves in the grand scheme of things. It was a revolution that occurred mainly in the seventeenth century. It was part of the birth of modern science, but it went far beyond science; it was part of the beginnings of Protestantism and the victory of natural theology over Scripture, but it went far beyond religion; it was part of the French Enlightenment and the Age of Reason. The extraterrestrial question, the question of whether our minds are necessarily unique in the universe, penetrates to the deepest roots of our culture and our identity as human beings.

The notion of other worlds goes back at least as far as the Greek atomists and their view that space is filled by an infinite number of similar atoms, all obeying the same natural laws. In such a philosophy, whatever happened on earth would have been repeated all through the cosmos. Aristotle, however, strongly disagreed with this picture. All things, according to Aristotle, were composed of five elements: earth, air, fire, water, and ether—and each element had its "natural place." The natural place of "earth" was at the center of the universe, and all earthlike particles anywhere in the universe would fall to that place. The natural place of ether was in the outermost heavens, where it made up the stars.

Water, air, and fire had intermediate locations. Aristotle's intellectual grip through the centuries was powerful. For one thing, his view of an earth-centered cosmos appeals strongly to common sense. Standing outside on a starry night, it's easy to believe that the universe revolves around us and that those distant points of light are made of some nonearthly material. Everything in its place. Aquinas made it his business to reconcile Christianity with Aristotelian philosophy whenever possible.

Aristotle aside, there were strong religious and emotional reasons for rejecting the possibility of other worlds. Only one earth is mentioned in Scripture, for example. Perhaps more importantly, the whole tone of the Bible suggests a comfortable, personal relationship between man and God. God watches over us. In medieval times, it was also widely believed that the universe was created principally for us and our use. Life is baffling as it is. Who wants to live out her days with dubious status, in a cosmos with uncertain purpose? People were not in a hurry to give these things up.

Even so, not everyone was able to wed faith and reason so peacefully as Aquinas. God's omnipotence and creative power would be diminished by being limited to a single world, pronounced the bishop of Paris in 1277. This was a forceful argument in favor of other worlds (essentially the same as Hesburgh's in the NASA book), and it appeared frequently. Theologically, it wasn't clear whether a plurality of worlds would enhance or diminish God's glory. For the next few hundred years, the possibility of other worlds was hotly contested among intellectuals.

In 1543, a critical new element was added to the debate, a scientific element. The *Revolutions of the Heavenly Spheres* of Copernicus was published, announcing that the astronomical data were best fit by placing the sun, rather than the earth, at the center of the solar system. Copernicus didn't speculate on whether earth's sister planets were earthlike, not to mention inhabited, but the writing was on the wall. This was the beginning of many contributions of science to the question of other worlds, as well as the beginning of modern science itself. In the next century and a half, Galileo's telescopic sightings of irregular mountains on the moon, Kepler's observations of the sudden appearance of a "new" star in the sky where none was before, and Newton's law of universal gravity would all become ammunition in support of the possibility of other worlds.

But it is hard to believe that these technical developments by themselves were the force that eventually turned people's opinions. What I find revealing in this regard is that many of the scientific arguments of this period wobbled on shaky ground, many smelled quaintly of old-fashioned human prejudice and egotism, and all of them invoked the Deity in some form or another. The obvious proving ground for the existence of extraterrestrial life was the moon, being closest at hand. To make lunar conditions more comfortable, Galileo (erroneously) hypothesized that the smooth periphery of the moon seen in his telescope was caused by a lunar atmosphere and that the dark spots were lunar oceans. The great German astronomer Johannes Kepler (erroneously) concluded that the shapes and arrangements of the lunar hollows were evidence of architecture by intelligent creatures—so similar in flavor to Percival Lowell's famous "observations" of artificial canals on Mars at the turn of our own century. Although a strong advocate of life on the moon, Kepler took pains to point out that, according to his calculations, our sun was the most luminous (and therefore the noblest) star in the Milky Way. This erroneous result may, perhaps, be understood in light of his comments in *Kepler's Conversation with Galileo's Sidereal Messenger,* 1610:

> . . . if there are globes in the heaven similar to our earth, do we vie with them over who occupies the better portion of the universe? For if their globes are nobler, we are not the noblest of rational creatures. Then how can all things be for man's sake? How can we be the masters of God's handiwork?

Kepler was straddling the fence. Other inhabited worlds in our solar system were all right, but our particular setup was still the best and the brightest. And the astronomer Thomas Wright, in his *Original Theory or New Hypothesis of the Universe* (1750), brandishes his authority as a scientist to support the plurality of worlds, but states up front that "the glory of the Divine Being of course must be the principal object in view" and uses that object in his construction of the cosmos. A lot more than science was at the bottom of these convictions.

During this same period, theology itself was changing dramatically, especially in England. It was increasingly believed that God revealed Himself more in His natural works than in Scripture. Nature was celebrated. This shift in view eventually showed up in the philosophy of Rousseau, in the nature poems of Coleridge and Wordsworth, and in the landscape paintings of Turner and Constable. It was also brought to bear on the question of other worlds. In 1638 the Protestant clergyman John Wilkins, who later became an Anglican bishop, bravely argued that the lack of mention of other worlds in the Bible did not forbid their existence. Some fifty years later, the

English theologian Richard Bentley carried the new "natural theology" to its ultimate implications for man:

> . . . we need not nor do not confine and determine the purposes of God in creating all mundane bodies, merely to human ends and uses. . . . All bodies were formed for the sake of intelligent minds: and as the Earth was principally designed for the being and service and contemplation of men; why may not all other planets be created for the like uses, each for their own inhabitants which have life and understanding?

And in Milton's *Paradise Lost* (1667), the archangel Raphael answers Adam's cosmological questions in this way:

> . . . other suns perhaps
> With their attendant Moons thou wilt descry . . .
> Stor'd in each Orb perhaps with some that live.
> For such vast room in Nature unpossesst
> By living Soul, . . . is obvious to dispute . . .

Theological considerations were appearing here in a different garb than before. God was still a powerful and good force, but something had changed underneath all the dressing.

As stated so well by Bentley, humankind was becoming humble—at least intellectually. A landmark in the new humility was Descartes' *Principles of Philosophy* (1644), the most comprehensive study of knowledge since Aristotle's, and an enormous influence on modern thought. In *Principles,* Descartes thinks about everything from the nature of thought

itself, to the five human senses, to the movement of
projectiles, the behavior of fluids, sunspots, the
mechanisms of tides, the nature of mind, and the
soul. Although this may not seem like the undertak-
ing of a modest man, Descartes places great stock
in that quality as a basis for reasoning. In the third
part of his work, before launching into cosmology,
Descartes warns us not to presume too much in understanding God's purposes and then suggests that
those purposes are most likely not all for our benefit.
Furthermore, and throughout Descartes' vast, nonan-
thropocentric cosmos, nature follows universal laws.
Nature is a single, mechanical system, in which fluids
and sunspots dance to the same rules here as they do
everywhere in the cosmos. Descartes' ideas drifted
like pipe smoke through the salons of Holland,
France, Germany, and England. Although the details
of his science were soon to be swept aside by Newton's *Principia,* Descartes' redefinition of man's place
in the universe had struck deeply and stuck. Descartes, so it seems, was underneath the comments of
Bentley as much as Copernicus was.

In 1686, Descartes' philosophy enjoyed perhaps its
most literate and widely read expression, in the classic
Conversations on the Plurality of Worlds by Bernard Le
Bovier de Fontenelle. Fontenelle—writer, philosopher, secretary of the French Academy of Sciences
for half a century, important figure in the French
Enlightenment—was unsurpassed in his ability to convey science to the general public. In *Conversations,*
Fontenelle meets a cultured lady for several evenings
of pleasant conversation. As they stroll through the
park, he unfolds before her the new universe of Copernicus and Descartes—in nontechnical, witty, and po-
etic language. It is a universe in which nature is like a

watch, and it is a universe not designed for our convenience. It is a universe in which inhabited planets orbit other suns. In Fontenelle's lifetime alone, *Conversations* went through twenty-eight editions. It was translated into English the year after its first publication and later into German. Other popular books with the same message soon appeared. As the eighteenth century got under way, the possibility of other worlds quietly slipped into Western culture.

So much for a brief history of a very big idea. Today, most of us have a modest view of our place in the universe, without thinking much about it. Our ancestors' bones look a lot like the bones of apes. We've seen pictures of our fragile planet taken from the moon. Rocks from space have crashed into our backyards.

The other day, I walked over to the house of a neighbor, a freshly ordained Episcopal priest, and asked her how she felt about extraterrestrials. They were fine by her. How would she react if we made contact with them tomorrow, I asked. She'd want to know their value system, she answered.

GRAVITATIONAL WAVES

We went up to Cambridge the other day to hear a lecture on gravitational waves, given at the Harvard-Smithsonian Center for Astrophysics by the physicist Rainer Weiss, of the Massachusetts Institute of Technology. The theory that gravitational waves exist was propounded by Einstein in 1916, but they have not yet been directly detected. Their effects are estimated to be incredibly tiny. Professor Weiss and his group, together with a dozen other such groups around the world, are betting their careers and some of the taxpayers' money that gravitational waves can eventually be found and will prove interesting when they are. In the lecture room, Professor Weiss was given a long introduction by Robert Vessot, a straightforward, nuts-and-bolts type of physicist, who remarked that he and Professor Weiss had been friends from way back —ever since they were young, rowdy bachelors—and that both are now in their fifties and have families. Then Professor Weiss got under way. Weiss, a high-strung, rumpled-looking man, is revered as a genius at experiments, and the audience of physicists and astronomers got very quiet.

Professor Weiss began by saying that a gravitational wave has only about a trillionth of a trillionth of a trillionth the strength of a radio wave, other things being equal. For that reason, only extremely violent events, like the explosion of a star, will produce enough gravitational waves to amount to anything. We can forget about trying to produce any ourselves. (Someone asked if a nuclear explosion would create noticeable gravitational waves, and Professor Weiss said he'd considered that and the answer was no.) What makes it desirable to detect gravitational waves is that they could give us an inside view of certain astronomical objects, which have so far been visible only from the outside. Radio waves and light waves created in the middle of stars are absorbed before they can escape into space; the waves we see come only from the surface. But stars would be practically transparent to gravitational waves. A gravitational wave created in the middle of a star would travel almost unimpeded from there to here, bringing news about its place of birth.

Professor Weiss then described a gravitational-wave detector that has been proposed by his group at M.I.T., in collaboration with a group at Caltech under the physicist Ronald Drever. Actually, two detectors would be constructed—one in Maine, near Columbia, and the other in California or Nevada—at a total cost of about sixty million dollars. Each would consist of three masses, weighing about a ton each, suspended inside an L-shaped tube four feet in diameter and five miles long. Laser beams would travel down the two legs of the tube, be reflected by mirrors fastened to the suspended masses, and gauge the motions of the masses as a gravitational wave passed by and jiggled them. This procedure would have to be carried out with great delicacy. According to the theory, a typical gravitational wave arriving from space would jostle the masses less than the width of the nucleus of an atom. (Someone in the audience asked if the detector in Maine wouldn't be shaken up much more than this by the pounding of the ocean, thirty miles away. Professor Weiss said he'd considered ocean waves and thought he would be able to take them into account, and, besides, his colleagues and graduate students were strongly in favor of the Maine site.) M.I.T., Caltech, and other groups have already built smaller, prototype detectors that demonstrate the main principles.

Throughout his talk on gravitational waves, Professor Weiss took their existence as a given. Some scientists, he said, were unbelievers until recently. About ten years ago, the physicist Joseph Taylor discovered a system of two stars that are orbiting each other and slowly spiralling together, and subsequent observation confirmed that the spiralling motion agrees perfectly with the predicted frictional effect of gravitational waves being generated by the two stars. Professor Weiss appeared extremely pleased with this confirmation, but he made it clear that he would have believed in gravitational waves anyway.

Professor Weiss seemed to us very evenhanded about comparing the M.I.T. experiment with related experiments elsewhere. As far as he is concerned, the search for gravitational waves is a community affair. "We need to get the Nobel Prize fever behind us, and work together on this thing," he said. Professor Weiss has been talking to scientists and to funding agencies since 1972 to gain support for the project. If the money comes through, the detectors could be ready in four or five years. "There are two groups of scientists, and the gravitational-wave experiments should satisfy both," he said. "There are the sharpshooters, who want answers to specific questions, like whether a gravitational wave travels at the speed of light, as predicted; and there are the others, who fire buckshot and hope to hit answers to questions they haven't thought of." Professor Weiss gleefully admitted to being in the second group himself.

A TELEGRAM
FROM CLARENCE

Not long ago, a curious telegram came my way
through the kindness of Owen Gingerich, astronomer
and historian of science at Harvard and the Smithsonian. The telegram is addressed to Dr. Harlow
Shapley, the late, great American astronomer, and,
according to Professor Gingerich, came fluttering out
of a large box of Shapley's old clippings and papers
sent him to sort through. It's stamped "1925 Jul 10
A.M. 3:08" and reads:

DISTINGUISHED COLLEAGUES OF YOURS HAVE
SUGGESTED YOU MIGHT BE WILLING TO COME TO
TESTIFY FOR DEFENSE AT DAYTON TENNESSEE
NEXT WEEK IN THE CASE OF STATE OF TENNES-
SEE VERSUS PROFESSOR SCOPES STOP WE OF THE
DEFENSE WOULD BE DELIGHTED TO ADD YOUR
AUTHORITY TO OUR POSITION STOP YOUR EX-
PENSES WILL BE PAID STOP WILL YOU WIRE ME
DIRECTLY AT DAYTON AND I WILL LET YOU
KNOW WHAT DAY YOU WILL BE NEEDED.
                          [SIGNED] CLARENCE.

For reasons unknown, Shapley never went to Dayton for the trial. Eight other reputable scientists did—including two zoologists, two geologists (one, Kirtley Mather, a devout Baptist), an anthropologist, a soil expert, an embryologist, and a psychologist—and described in written statements how the surface of the earth had shifted and changed, how time can be reckoned by counting the light and dark bands of clay laid down as the great glaciers retreated, how the oldest rocks date back at least 100 million years and reveal no signs of life, while rocks of lesser age contain fossils first of simple plants and later of simple animals without backbones, how the body structures of monkeys and apes and humans appear similar, how man springs from a lineage of decreasing bony eyebrow ridges and increasing brain capacity, and how a human embryo develops gill slits like a fish, a three-chambered heart like an amphibian, and a tail.

Judge John T. Raulston ruled to exclude the scientists' expert testimonies, on the grounds that the question was whether Scopes had broken the law, rather than a contest between the biblical account of creation and the theory of evolution. But on Monday, July 20, ten days into the trial, he allowed their statements to be read for the record, with the jurors absent. The courtroom was packed. Announcing that cracks had developed in the ceiling of the room below, Raulston judiciously herded the court and masses of interested spectators outdoors to safety to hear the conclusion of the scientific remarks. Two thousand people converged on the courthouse lawn that hot afternoon and listened, while boys worked the crowd selling soda pop, to a story that was better than could be found in any book.

Had he answered Clarence Darrow's call for help, Shapley might have offered astronomical evidence that the stars were shining long before 4004 B.C., the date of Creation as computed by Archbishop Ussher. He might also have explained that our sun is simply one of billions of suns in the galaxy and that our solar system is located not at the center but well off to one side. Shapley himself discovered this last astronomical item in 1917, at the age of thirty-two. Shapley's conclusions, like Darwin's before, were supported by collecting a great deal of data and, as far as possible, letting the facts speak for themselves.

Some of the facts had been collected by others. It was the Italian monk Giordano Bruno, a man willing to take on the church and Aristotle, a man burned at the stake in 1600 for his unpopular thinking, who suggested that the stars in the heavens might be suns like our own. Isaac Newton, in the next century, followed Bruno's lead and estimated the distance to nearby stars on the assumption they were emitting about the same amount of light as our sun. Like a light bulb, a star appears brighter the closer it is, and this relation can be quantified to measure distance to the bulb, if you know or assume its wattage (luminosity). Bruno's hypothesis and Newton's assumption were confirmed in 1838, when the distances to the nearby stars were accurately measured and translated into stellar luminosities that roughly matched the sun's (400 trillion trillion watts if you're counting). Not all stars in the sky have the same luminosity, but our own star, the sun, is rather typical. Long before these specifics, however, people knew that the stars formed an organized structure which makes up our galaxy and appears on a clear night as a misty band of light overhead. Our galaxy is called the Milky Way. There are others. Galileo, pointing the newly invented telescope toward the heavens in 1610, noted that the Milky Way is in fact a congregation of individual stars. In 1785, the British astronomer William Herschel showed that the Milky Way takes the shape of a grindstone.

In the early 1900s, the size of the grindstone and our position in it were measured by the Dutch astronomer Jacobus Cornelius Kapteyn. His erroneous results, which held sway until the work of Shapley, were that the Milky Way is some 10,000 light years in diameter and is centered on our sun. Kapteyn did not properly take into account interstellar dust, which partially absorbs starlight and mocks the effect of a decreasing density of stars in all directions away from us. Given this apparent thinning of stars in every direction, Kapteyn couldn't be blamed for figuring he was in the middle of things.

Harlow Shapley was a tenacious man. Take his fascination with ants, for example. During his epic work on the Milky Way, he would study the stars by night and the ants by day. This happened near the top of Mount Wilson in southern California, where the sixty-inch telescope was perched. As it was a nine-mile hike up the mountain, Shapley and the other astronomers would hole up on the summit for days or weeks at a time before coming down, observing at night and sleeping during the day. But a person couldn't use up *all* the daylight hours sleeping. One day Shapley was idly watching a stream of ants running along a concrete wall when he noticed that the ants slowed down when they entered the shade of some manzanita bushes. Wondering about this, he secured a thermometer and a stopwatch and established an unofficial observing station. To run so many centimeters at such and such a temperature took so many seconds. Shapley had, singlehandedly, discovered the thermokinetics of ants—the higher the temperature, the faster they run —and, after carefully quantifying it all, boasted that he could reckon the temperature to within one degree by observing half a dozen ants in his speed traps.

Now to Shapley's nighttime activities at Mount Wilson. Ever since the Greeks began gauging the distances to the moon and the sun, a major aim of astronomy has been to find out how far away are the heavenly bodies. This task has an unpleasant similarity to staring at a photograph of totally foreign objects and trying to figure out their actual sizes. You have to start with something you know, if you can find it, and then compare. Shapley was interested in measuring the distances to the far-off stars in globular clusters. A globular cluster is a shining ball of about a million stars, all orbiting each other under their mutual gravity, beautiful to behold. A globular cluster is an intermediate-sized astronomical object, far larger than a single star and far smaller than an entire galaxy. There are some one hundred of them in our galaxy. Many globular clusters contain one or more peculiar stars called Cepheids, which provided the key to Shapley's discoveries. Find the distance to a Cepheid star in a globular cluster and you have found the distance to that cluster, just as you know the distance to a far-off ship if you know the distance to one of its passengers.

Cepheids are among the class of so-called variable stars, whose brightnesses vary in a periodic manner. There is strong reason to believe these light variations are caused by material pulsations of the star as a whole, bodily breathinglike contractions and expansions. In 1912, Henrietta Leavitt at Harvard showed that a relation exists between the average luminosity of variable stars and the cycle time (period) of their pulsations. Also at Harvard, Solon Bailey had detected a number of Cepheid variable stars in globular clusters and suggested to the young Shapley, on his way from graduate school at Princeton to

Mount Wilson, that this was an area of study worth looking into. Shapley set to work discovering Cepheid stars and graphing the luminosity versus period of 230 of them, ranging in period from five hours to one hundred days. This was an immensely tedious project, requiring numerous photographic plates of each star in order to detect its brightening and dimming over time. To embroil himself in such a long and painstaking labor, an astronomer would have to be highly convinced of its ultimate payoff.

Even with Shapley's graph, one thing was missing. The graph gave only the *relative* luminosities of Cepheid stars of different periods. To put everything into *absolute* terms, to firmly identify the one object in the photograph against which all else could be scaled, Shapley directly measured the luminosities of some nearby Cepheids. Determining the distance to a globular cluster was then straightforward: find a Cepheid in it; measure the Cepheid's period of variation, and by comparison with the graph, thus infer its luminosity; then measure how bright it appears and thus conclude how far away it must be.

After Shapley had figured out the distances to the globular clusters in the Milky Way, he could plot their arrangement in space. He didn't expect what he found. The clusters mapped out an enormous sphere whose diameter was thirty times larger than that calculated by Kapteyn and whose center was off in the direction of Sagittarius, 60,000 light years away. (More modern measurements indicate that our galaxy is only one-third as large as Shapley's value and that our sun is about two-thirds the way out from the center.) Evidently, the Milky Way has a large spherical halo in addition to its central disk. More important, the center is not us. As Shapley commented, "The solar system is off-center and consequently man is too . . . he is incidental."

When Shapley got the telegram from Clarence he may have been sitting at his famous rotating desk at the Harvard College Observatory, where he was director from 1921 to 1952. The observatory, with its brick buildings and modest telescopic domes, sits on a hill in Cambridge, Massachusetts, an average-sized town on one of nine planets revolving about an average star. Last year, 1985 according to the local standards, marked 100 years since Shapley's birth, 512 since Copernicus', 176 since Darwin's, 129 since Freud's— all human beings who did extraordinary work to show how ordinary are human beings.

THE ORIGIN OF
THE UNIVERSE

For the last couple of decades, physicists have been pushing their theories of matter and energy backward in time, closer and closer to the primal explosion that started the universe, some 10 billion years ago. Indeed, it is common these days for scientists to hold forth on "The Origin of the Universe." When you go to such lectures, however, you soon learn that you're not getting The Origin, but a billionth or a trillionth of a second later.

And so it was that I casually took my seat at yet another lecture titled "The Origin of the Universe," given at Harvard in the spring of 1984. The lecturer was the British scientist Stephen Hawking. The hall was packed. Hawking, then forty-two, has become one of the seminal theoretical physicists of our time. He has also suffered for years from a worsening motor neuron disease, which has ravaged his body but spared his mind. On this afternoon, as Hawking sat in his wheelchair, laboring to utter a series of sounds that were translated into words by a student, I gradually realized what I was hearing: Hawking had traveled back the whole distance. For the first time, a preeminent scientist was tackling the *initial* condition of the universe—not a split second after the Big Bang, as I'd heard about before, but the very beginning, the instant of creation, the pristine pattern of matter and energy that would later form atoms and galaxies and planets.

In any other circumstance, physicists would hardly think twice about discussing initial conditions. The "initial conditions" along with the "laws" are the two essential parts of every model of nature. The initial conditions tell how the particles and forces of nature are arranged at the beginning of an experiment. The laws tell what happens next. Any predictions rest on both parts. Set a pendulum swinging, for example, and its motion will be determined by the initial height where your hand let go, as well as by the laws of gravity and mechanics. But Hawking's pendulum is the entire universe. And he is attempting to reason out what theologians and scientists alike have previously assumed as a given. He is attempting to *calculate* where the hand let go. Hawking's equations for the initial state of the universe, together with the laws of nature, could predict the complete outcome of the universe. They could tell us whether our universe will expand forever or reach a maximum size and then collapse. They might explain the existence of planets, or of time.

How could anyone know whether Hawking's equations are right? Could the human mind even grasp Creation? And, just as puzzling, how did science arrive at such outrageous self-confidence? I asked these questions to myself and a dazed colleague as we wandered out of the lecture hall and away from the campus, walking past cars and children in mittens.

Physicists today are not modest—and with some

reason. Just in this century, they have discovered and successfully tested a new law for gravity, a theory for the strong nuclear force, and a unified theory of the electromagnetic and weak nuclear forces. They have proposed further laws that might unify all of the forces of nature. Physicists have demonstrated that time doesn't flow at a uniform rate and that sub-atomic particles seem to occupy several places at once. These victories, often in territory far removed from human sensory perception, have created a strong sense of confidence.

In many of the more heady advances, theory has outdistanced observation, let alone application. For example, the unified theory of the electromagnetic and weak nuclear forces, developed in the 1960s, predicted the existence of new particles that weren't discovered in the laboratory until the 1980s. Super-dense stars, fifteen miles in diameter, were predicted in the 1930s—more than thirty years before they were first observed in space, thousands of light years from earth. Einstein's general theory of relativity predicted that a ray of starlight passing near the sun should be deflected five ten-thousandths of a degree by the sun's gravity. When a novel experiment confirmed this minuscule effect several years later and Einstein seemed blasé, a student asked him what he would have done if his prediction had been refuted. He answered that then he would have been sorry for the dear Lord, because "the theory *is* correct."

With such confidence, physicists have grown accustomed to extrapolating their theories to situations that could not possibly be witnessed by human beings. Hawking's work on the beginning of the universe is an extreme example of this kind. In the evolution of the universe as a whole, gravity is the dominant force to be reckoned with. Hawking has extrapolated Einstein's theory of gravity back to an epoch that was not simply prior to life, but prior to atoms. Stranger still, the early universe was of such high density that its entire contents, including the geometry of space itself, behaved in the hazy, hard-to-pin-down manner of subatomic particles. The methodology needed to describe such behavior is called quantum mechanics, and the application of this methodology to gravity is called quantum gravity. According to the theory of quantum gravity, it was possible for the entire universe to appear out of nothing.

Hawking has mathematically investigated the kind of universe that could have appeared out of nothing. Would the infant universe be finite or infinite in extent? Would it curve in on itself? Would it look the same in all directions? Would it be expanding rapidly or slowly? The answers to these questions are buried in a difficult equation. That equation will likely take a long time to solve, an even longer time to test against its predictions, and it may be plain wrong. Nevertheless, it reflects an extreme confidence in the power of human reason to reveal the natural world. Hawking, like Darwin, has ventured into regions previously forbidden to human beings, and to scientists in particular. Hawking's work, right or wrong, is a celebration of human power and entitlement to knowledge.

Men and women have always longed to understand and control their world, but they have constantly met with obstacles. In different ages and cultures, they have tried different means to clear their path—magic in primitive cultures, religion and science in more evolved societies. Primitive man believes in his power to control nature and other men through magic. He believes he can make rain by climbing up a fir tree and drumming a bowl to imitate thunder. He believes he can bring a cool breeze by wrapping a horsehair around a stick and waving it in the air. But with growing experience, man realizes that these methods have their limitations. Rain and cool breezes don't always come when requested. At this stage of development, as the anthropologist Sir James Frazer says in *The Golden Bough,* man stops relying on himself and throws himself at the mercy of higher beings. Thus begins religion. And a surrender of personal power. My rabbi once told me that man has always made of God what he wished to be himself.

But, as man increases his knowledge, this new reckoning with the universe also needs revision. For the gods, reflecting the ignorance and superstition of man, have often been given human personality along with their powers. The gods get drunk, as the Babylonian deities did on the night before Marduk went out to do battle against Chaos. They are jealous and spiteful, like Hera, who destroyed the Trojan race because she had placed runner-up in a beauty contest judged by a Trojan. If natural phenomena are controlled by such gods, then those phenomena should be subject to whim and to passion. Yet the more man studies nature, the more he finds evidence for regular laws. Seasons repeat, stars move on course, and stones fall predictably. The study of these regularities marks the method of science. Through science, man regains much of his primitive confidence in self-power, with control now replaced by knowledge. Knowledge is power. Man might not be able to control the weather, but he can try to predict it.

With the beginning of modern science in Europe, people measured eclipses, they dissected cadavers, they observed mountains on the moon with the new telescopes, they peered at lake water through microscopes, they studied magnets and electricity. Copernicus declared that the earth went around the sun. Paracelsus announced that disease was caused by agents outside of the body, not by internal humors. Galileo pointed out that moving bodies maintained their motion unless acted on by external forces.

Yet flowing deep through human culture remained the idea that some areas of understanding were off limits or beyond mortal grasp. Adam and Eve were punished because they ate from the forbidden tree of knowledge, which opened their eyes and made them "as gods." In *Paradise Lost,* Adam asks the angel Raphael to explain Creation. Raphael reveals a little, and then says:

> . . . the rest
> From Man or Angel the great Architect
> Did wisely to conceal, and not divulge
> His secrets to be scann'd by them who ought
> Rather admire. . . .

Doctor Faustus approached other authorities for knowledge and had to pay with his soul. People also questioned to what extent the universe was subject to human rationality. Descartes had likened the world to a giant machine, but many viewed such reductions as threats to the power of God. In his Condemnation of 1277, the Bishop of Paris made it clear that no amount of human logic could hinder God's freedom to do what He wanted. Even Isaac Newton, master logician and reductionist, surveyor of all natural phenomena, comes to the end of *The Principia,* the General Scholium, where he lets down his hair and confesses that the synchronized performance of moons and planets could never be explained by "mere mechanical causes," but requires "the counsel and dominion of an intelligent and powerful Being." Furthermore, it would be impossible for mortal man to fathom the art of that divine balancing act: "As a blind man has no idea of colors, so we have no idea of the manner by which the all-wise God perceives and understands all things." Newton, both scientist and believer, was caught between his own power of calculation and the unknowable power of God.

But the unknowable continued to beckon, and man, although fearful of lifting the veils, was still driven to try. After Newton, a great debate arose over whether the solar system could be explained on a rational basis. Similar debates echoed through later centuries. In the eighteenth and nineteenth centuries, geologists argued over whether changes in the earth came about through gradual transformations, obeying natural law, or through sudden catastrophes, ordered by a tampering God. The mode of thinking in the late 1800s, just before Madame Curie found that the sacred atom could be splintered, was described by Henry Adams in this way: ". . . since Bacon and Newton, English thought had gone on impatiently protesting that no one must try to know the unknowable at the same time that every one went on thinking about it."

For Henry Adams, the unknowable that became known was the atom. For modern biologists, it is the structure of DNA and possibly the creation of life. For modern astronomers, it is the distance to the galaxies and the shape of the cosmos; for modern physicists, perhaps the grand unified force and the birth of the universe. Layer by layer, the unknowable has been peeled away, and examined, and made rational. Scientists today, humbled by their dwindling size in the cosmos but emboldened by their success at adjusting, have staked out all of the physical universe as their rightful territory. And they intend to let their theories and equations take them to places they cannot go with their bodies. In the introduction to one of his recent papers, Hawking says: "Many people would claim that the [initial] conditions [of the universe] are not part of physics but belong to metaphysics or religion. They would claim that nature had complete freedom to start the universe off any way it wanted. . . . Yet all the evidence is that [the universe] evolves in a regular way according to certain laws. It would therefore seem reasonable to suppose that there are also laws governing the [initial] conditions."

To me, Hawking's work, although strikingly bold, is a natural extension of what science has been doing for the last five hundred years. But the question remains: After physics has reduced the birth of the universe to an equation, is there room left for God? I asked this to a colleague who has done significant calculations on the origin of the universe and is also a devout believer in God. He answered that while physics can describe what is created, Creation itself lies outside physics. But with your equations, I said, you're not giving God any freedom. And he answered, "But that's His choice."

TINY PATTERNS

On a spring day in 1936, shivering in the bitter cold of his laboratory, Dr. Ukichiro Nakaya of Hokkaido University created the first man-made snowflake. The trick, it seems, was a thoroughly dry rabbit's hair. Mounted inside an ingenious container of rising warm water vapor and falling cool air, the hair sprouted ice crystals along its length and got the snowflake started. Up in the clouds, tiny bits of dust serve the same purpose.

Twenty-seven years later, fully ignorant of my predecessors in snowflake research, I bounded out-of-doors in a fresh snowfall with my new microscope and spent several hours catching snowflakes on cold glass slides and peering at them under magnification. Like other fourteen-year-olds, I was filled with awe and questions at the same time. How could each snowflake be shaped so differently? Could this keep up, with countless trillions of them dropping from the sky? And yet, despite this diversity, every flake had nearly perfect six-sided symmetry, repeating each fragile branching and va-

gary six times around. How did a snowflake know to
do that?

These questions had, of course, been raised before.
The German astronomer Johannes Kepler was one of
the first observers to document the six-sided symmetry
of snowflakes, spelled out and pondered in a little
essay of 1611, "A New Year's Gift, or On the Six-
Cornered Snowflake." All without benefit of a microscope, which was barely known at the time. Kepler,
having spent many years mathematically taking apart
the patterns of planetary orbits, was especially sensitive to questions of design. In his paper, he reasoned
through several theories for how solids, in general, are
structured. Kepler realized that geometrical truths
alone will bring about certain regularities when identical objects are packed together. For example, if you
stuff golf balls into a large box, two special arrangements maximize the number of balls that will fit: one
with a four-sided symmetry and one with a six-sided
symmetry. (You have to ignore appearances near the
edges, which depend on the shape of your particular
box.) The moral is that haphazard shuffling always
leaves you with more balls left over than necessary.

For pure solids, like ice, molecules take the place of
the golf balls. It's a little different at the microscopic
level, because packed molecules don't actually touch,
but, again, only certain symmetrical arrangements of
the molecules are most efficient at construction. Nature, left to her own, usually follows the path of economy. The problem is that snowflakes don't form side
by side in boxes. The internal six-fold symmetry at the
very core of each snow crystal, where a mere few
thousand molecules are shouldered together in a tiny
chunk of ice, can indeed be explained in terms of
geometrical economy and the particulars of the water

molecule. But as a snow crystal grows, each of its six arms stretches outward, without touching its neighbors or other snowflakes, and eventually recruits a million trillion molecules in the process, holding the six-sided symmetry all the while. Something besides geometry is going on here.

No less astonishing, but closer perhaps to being explained, are the endlessly inventive patterns of snowflakes, reminding me of that haunting last line of *The Origin of Species:* ". . . from so simple a beginning endless forms most beautiful and most wonderful have been, and are being, evolved." People have gone to great lengths to catalog the various species of snowflakes. One Wilson A. Bentley, later known affectionately as The Snowflake Man, spent winter after winter in a small open shed in northern Vermont capturing and photographing snowflakes through a microscope. A small, birdlike man who never married, Bentley was enthralled with his snowflakes, worked around the same shed for forty years, and netted nearly six thousand photomicrographs, all told. His classic book *Snow Crystals,* published shortly before his death in 1931, helped motivate the modern division of crystal shapes into plates, stellars, columns, and needles. Bentley himself didn't squander his time on theory.

My colleagues in atmospheric physics don't seem surprised to find so much variation within the four basic shapes. Snow crystals grow by the condensation of cold water onto a seed of dust or ice, as Nakaya demonstrated fifty years ago. It is also known that the growth pattern depends sensitively on the temperature and humidity of the surrounding air. Needles form at air temperatures between 23 and 27 degrees Fahrenheit, hollow hexagonal columns between 18 and 23 degrees. Higher concentrations of water vapor favor more exaggerated shapes, like ornately branched crystals (called dendrites) instead of flat plates. And so on. Aloft for perhaps two hours in its journey to earth, buffeted by winds in all directions, a growing snowflake lives through a turbulent history of local weather conditions—each minute deviation, each different trajectory causing a different montage in the end.

Lately, scientists have taken to growing snowflakes on computers. The way this works is they write down some physics equations describing all the relevant processes of formation (like the release of latent heat when a liquid freezes), simplify as much as they dare (like playing in two dimensions instead of three), specify the environment surrounding the nascent snowflake, and hand the whole thing to a computer to grind out numbers and draw pictures. The hope, of course, is that promising shapes will appear on the screen. A leading scientist in this kind of work is James Langer at the Institute for Theoretical Physics in Santa Barbara, California. Langer and coworkers have gotten sufficiently realistic pictures to cautiously claim that they know what they're doing. A bit of ice poking out from their theoretical snowflake loses heat to the surrounding subfreezing air more rapidly than the rest of the flake, causing the hypothetical air to condense into ice more rapidly at that point. The stray bit, enlarged, pokes out more. This leads to even more growth and temperamental branching—just what the doctor ordered to explain diversity of form. More profound and not yet understood, the complex equations for all this seem to predict certain preferred patterns of growth. As a consequence, each of the nearly identical arms of the initial ice crystal will slip into the same pattern of growth at every moment, perpetuating the six-sided symmetry. The scientific insight, and art, in such computer simulations lies in cutting down the problem to a manageable size without leaving out vital effects.

Langer is getting warm, but I bet he's in for more surprises. Real snow has magic.

One night last winter, a giant snowstorm hit Concord, Massachusetts, dropping nearly a foot of dry white powder. My family was up early the next morning. Outside, the misshapen world looked invitingly unfamiliar. We bundled up in scarves and mittens, borrowed a great wooden toboggan from the neighbors, and headed out to a sloping field nearby. By my estimates, there must have been a million billion snowflakes in that field, with more sprinkling down, each a tiny mirror in the sun and different from all the others. We climbed to the top of the hill and installed ourselves in the toboggan, me in the front, my wife bringing up the rear, and our three-year-old daughter wedged in between. No one else was there. We pushed off. As we gained speed, a fine white mist came flying by, tingling our faces and sparkling with tiny patterns. Aside from my daughter's giggles, the only other sound was "the sweep of easy wind and downy flake."

WALDEN

I work not far from Walden Pond and recently paid a visit to the place. An extra hour had somehow slipped into my schedule, and I needed a walk in the woods.

These days an hour of quiet reflection is not to be tackled without preparation. To capture my unfettered thoughts before they floated out of reach, I took along my miniature tape recorder, something called a Realistic Minisette II, about the size of a pack of cigarettes. Unfortunately, a trial run at the last minute revealed the thing was busted, so I had to settle for pencil and paper. Actually, taking notes the old way has a soothing effect, once you get the hang of it.

I arrived at the pond about three o'clock on a hazy afternoon in May. I had to leave my car in a large cement parking lot on the other side of Route 126, a brisk thoroughfare, and take my chances crossing the street on foot. Then I trekked down a sloping hill to a well-worn footpath at water's edge. It being a weekday, I had the place almost to myself. Out on the pond there was a single boat, a red canoe, with a man and boy in it fishing. From time to time, the sun would glint from their metal rods, briefly calling attention to their otherwise muted and motionless forms. Possibly they were father and son, happily reeling in summer on their invisible lines. They had no other appointments; they were settled for the season. Now that seems like a sensible approach to keeping time. I glanced down nervously at my digital watch and moved on.

It's about two miles around the pond, longer if you take side trips into the woods. The water was high, on account of heavy rains, and a mallard floated indifferently beside a partly submerged tree. Also underwater were portions of the footpath, requiring agreeable detours through the foliage. One enormous tree lay mortally wounded on the forest floor, its upper leaves still green, its base in splinters.

Gradually I made my way to the site of Thoreau's house, nestled in the woods on the north side of the pond. Overhead, a flock of sparrows darted by and lighted noiselessly atop a maple. A mild breeze trickled softly through the pines. Standing among the stone posts that mark the house site I could hear little else, save for the steady grumbles and buzzes of cars and trucks rolling down the road at great speed a hundred yards away.

Ah, civilization. I wonder what Thoreau would have said about our modern fascination with speed and precision. We're certainly going somewhere fast, but we haven't the time to take a look where. A hundred fifty years ago Thoreau said he rarely met anyone who had built his own house. Today, Thoreau might have trouble finding people who can repair their own bicycles. Last week the transactions in a busy store that I frequent came to a standstill when the computer that keeps accounts got ill and no one knew how to mend it. To his credit, one customer boldly took the thing apart and fiddled for some time with the thermal printer, which happened to be his specialty. The rest of us spectators, increasing our numbers by the minute, were mightily impressed at this unexpected display of expertise, but it turned out the thermal printer wasn't the problem after all.

Carried along by the wave of progress, we have specialized to the point where we can hardly pronounce the names of the professions we depend on, much less manage our own affairs. Academic journals have subdivided and multiplied like rabbits. Subdivide and conquer. Twenty years ago I was a high school student and could read through most of *Scientific American;* today I am a professional physicist and can scarcely make out a word of the articles on biology. This seems to me a dangerous situation. We can either trust our lives to specialists or learn to make do with less.

I walked on. A man in short sleeves sat at the edge of the pond, fishing, spellbound by the play of sun on water. I hated to disturb him but wanted to talk. "Catch anything?" I asked.

"Not yet."

"What're you fishing for?"

"Anything that comes along." Apparently, this fellow didn't think he needed special lures for trout, other lures for bass, and so on, depending on the brand of fish, the weather, the time of day. I suppose he had the whole day off.

I walked on deliberately, nearly tripping over a beer can in my path. Loosen up, I told myself. The wood pencil in my breast pocket felt honest and comfortable. We love our modern technology, but we keep a warm spot in our hearts for our homespun past. Candles remind us of life before electricity, horse-drawn carriages bring back the days before gasoline engines. There is a welcome ease in the imprecise edges of a candle's light, the slower forms of transportation. I wish I knew which of today's items are going out of style, so that I might treasure them all the more. Handwritten letters will likely be replaced all too soon by crisp computer messages. Movies in public theaters may die out in favor of home video cassette systems. Keys may become antiques, by virtue of electronic locks that open when spoken to properly. I gingerly handled the keys in my pocket, stooped and picked some violets for my daughter.

At precisely three-forty, southwest of the pond, the Boston and Maine railroad thundered by on its way to South Acton. Perched on a bluff, the tracks run beside the pond for nearly a quarter of a mile. When my car is in the shop for a week, as happens now and then, I take the train to work. The first day or two I am in a prickly humor and brood about the inconvenient schedule, the frequent stops. Later in the week I learn the faces of the other passengers and become one of them, contented with my mode of travel, pleased to have some time to read. How readily we adapt to new circumstances when everyone around us is taking the same ride. Only when we look out the window and see another fellow pass us in a faster vehicle do we feel deprived. Despite constant improvement in material goods, we react to relatives, not to absolutes. If this were not the case, everyone living before the middle 1800s, before electric power and internal combustion, would have suffered incalculably for lack of televisions and dishwashers and automobiles. Yet many of these poor souls went about their business cheerfully, judging by the books they wrote. I suspect we could get by handily on less, once we readjusted, but we continue to channel much of our wealth and energy into technology and tout our new gadgets as a sign of progress.

My hour was up, and I had rounded the pond. I tallied the projected expenses for my outing:

Gasoline	$0.70
New batteries for my Minisette	1.20
A Diet Coke on the way back	0.50
Total	$2.40

All in all, it was a frugal visit, especially in terms of today's prices. I headed back to the parking lot in a thoughtful mood, noting the large sign that warns admirers of Thoreau not to do so while intoxicated.

On the highway back to work, doing fifty in my Toyota, I was nearly run off the road by an impatient LeMans driven by a teenager. I let him pass. I was enjoying the afternoon.

TO CLEAVE AN ATOM

In the spring of 1962 our family built a fallout shelter in the backyard. The President of the United States had been coming on the television set, pointing his finger at us, and telling us to go out and build a shelter. Some months earlier the government had distributed 25 million copies of a booklet called *Fallout Protection: What to Know and Do About Nuclear Attack.* I was fourteen and terrified that I would not live to be fifteen, and it was my pleading each night at the dinner table, as my three younger brothers sat quietly, that convinced my parents to dig up the backyard and put in a bomb shelter. It cost $3,000, exactly the price of the "H-Bomb Hideaway" featured in *Life* in 1955. The thing was finished just in time for the Cuban missile crisis.

A short-legged man who loved hiking created the first man-made nuclear chain reaction, on December 2, 1942, in a disused squash court at the University of Chicago. His name was Enrico Fermi. In Fermi's chain reaction, a subatomic particle called a neutron hits the nucleus of a uranium atom, cleaving it in two and releasing energy in the process. A uranium nucleus has quite a few neutrons of its own, and, after the split, a few of these go flying off individually along with the two main fission fragments. Each of the spawned neutrons eventually strikes a fresh uranium nucleus, splitting it in half, releasing more energy and more neutrons, and the activity rapidly multiplies, going faster and faster. The uranium nuclei are like a lot of cocked mousetraps on the floor, each loaded with several Ping-Pong balls waiting to jump into the air when the spring is triggered. Toss a single ball into the middle to get the thing started, and soon Ping-Pong balls will be zinging everywhere. Fermi kept his chain reaction from getting out of hand by constantly removing some of the neutrons, just as the frantic release of the mousetraps can be slowed by catching some of the balls in midair before they land on cocked traps. Fermi was almost unique in twentieth-century physics for being superb in both theory and experiment. He had, with others, conceived of nuclear chain reactions in early 1939. The whole idea of fission was only a few months old at the time.

Before 1938, everyone believed that atomic nuclei remained more or less whole, with the nuclei of some elements gradually disintegrating, a few small bits at a time. The emission of these bits is called radioactivity. Antoine-Henri Becquerel, a French physicist, first discovered radioactivity from uranium in 1896, and, soon after, the husband-and-wife team of Pierre and Marie Curie observed it from another element, radium, which lost weight little by little as it hurled out tiny particles.

In the early 1900s, scientists didn't know where in the atom radioactivity originated. Atoms were pictured as solid spheres of evenly distributed positive electrical charge, embedded with negatively charged particles called electrons. The electron, discovered in 1897, was clearly a subatomic particle. Its existence already contradicted the old Greek notion that the atom was indivisible. But the details of an atom's innards were largely unknown. Then, in a brilliantly straightforward experiment in 1911, Ernest Rutherford discovered the atomic nucleus. Rutherford fired subatomic particles at a sheet of gold. The projectiles he used were alpha particles, found by the Curies in their studies of radioactivity and known to be about one-fiftieth the weight of gold atoms. If the positive charge in an atom were thinly scattered throughout its volume, as believed, then the alpha particles should have met little resistance in passing through the target gold atoms. But some bounced straight back, apparently having struck something highly concentrated. What Rutherford had discovered was that the atom is mostly empty space, with a very tiny center of positive charge, about which the electrons orbit at great distance. The dense center, the nucleus of the atom, contains all of the atom's positive charge and more than 99.9 percent of its weight. It is roughly a hundred thousand times smaller than the atom as a whole. The booming-voiced Lord Rutherford strongly preferred simple, rough-and-ready experiments, and this was surely one. He also had an excellent nose for making predictions. His experiments had shown that the atom's positively charged particles, called protons, reside in the central nucleus. Rutherford went on to predict correctly that protons share their nuclear living quarters with other, uncharged particles, later called neutrons.

One of Rutherford's collaborators from 1901 to 1903 was a man named Frederick Soddy, who later won the Nobel Prize in chemistry. They worked together on radioactivity. Soddy was impressed by the energy emerging from the depths of the atom. As early as 1903, he commented in the *Times Literary Supplement* on the latent internal energy of the atom and, in 1906, wrote elsewhere that there must be peaceful benefits for society, given the key to "unlock this great store of energy." Soddy had unusual foresight. So did H. G. Wells, who stayed well abreast of scientific developments and paid close attention to the remarks of such men as Soddy. Wells, however, made darker forecasts. In 1914 he published a lesser-known novel, *The World Set Free,* describing a world war in the 1950s in which each of the world's great cities are destroyed by a few "atomic bombs" the size of beach balls.

In many ways, the discovery of nuclear fission got under way in 1934. That was the year that Irène Curie, daughter of Marie and Pierre, and her husband, Frédéric Joliot, discovered "artificial" radioactivity. Before then, all radioactive substances had been gathered from minerals and ores. Joliot and Curie found they could *create* radioactive elements by bombarding non-radioactive ones with alpha particles. Apparently, certain stable atomic nuclei, content to sit quietly forever, could be rendered unstable if they were obliged to swallow additional subatomic particles. The forcibly engorged atomic nuclei, in an agitated state, began spewing out little pieces of themselves, just as in "natural" radioactivity. Enrico Fermi, then working in Rome, immediately took his lead from the Joliot-Curie work but decided to see if neutrons rather than alpha particles could be used to produce radioactive nuclei. Alpha particles are positively charged and therefore partly repelled by the positively charged nucleus, but the uncharged neutrons, Fermi reasoned, would have an easier time making their way into the nucleus.

When these experiments proved successful, Fermi bombarded the massive uranium nucleus, containing over two hundred neutrons and protons, to see what would happen. He automatically assumed, as did others, that neutron bombardment of uranium would create nuclei close in weight to uranium. Then, in late 1938, the meticulous radiochemists Otto Hahn and Fritz Strassmann found in the remnants of bombarded uranium some barium—an element that weighs about half as much as uranium. There had been no barium in their sample to begin with. Apparently some uranium nuclei had been cut in two.

In December 1938, Hahn sent a letter describing his curious results to Lise Meitner, his coworker of thirty years. Meitner had been a respected and much loved physicist at the Kaiser Wilhelm Institute in Germany, but she was Jewish and had fled to Sweden five months earlier. At Christmas her nephew, physicist Otto Frisch, happened to pay her a visit and described the encounter: "There, in a small hotel in Kungälv near Göteborg, I found her at breakfast brooding over a letter from Hahn. I was skeptical about the contents—that barium was formed from uranium by neutrons—but she kept on with it. We walked up and down in the snow."

During their walk, Frisch and his aunt puzzled over how a single, slowly moving neutron could split in half an enormous uranium nucleus. It was well known that the protons and neutrons in an atomic nucleus are held together by strongly attractive forces—otherwise the electrical repulsion of the protons for each other would send them flying away. How could so many attractive bonds be broken by a single neutron? Frisch and Meitner realized that the answer lay in an idea put forth by the master Danish physicist Niels Bohr. In 1936, Bohr had suggested that the particles in an

atomic nucleus behave in a collective way, analogously
to a drop of liquid. Frisch and Meitner reasoned that
if such a drop could be slightly deformed from a
spherical shape, the repulsive forces of the protons
would begin to win out over the other, attractive
forces. The attractive nuclear force between two nuclear particles weakens very rapidly as their separation
increases, while the repulsive electrical force weakens
far more slowly. Flatten a sphere of particles and each
particle, on average, gets further away from its neighbors. Flatten it enough and the repulsive forces domi-
nate, splitting it in two and sending the two halves
flying apart at great speed. Frisch and Meitner calculated that the uranium nucleus was very fragile in
terms of these deformations and that a small kick from
a diminutive neutron might send it over the brink.
According to their figures, the energy release should
be enormous. Frisch went back to Copenhagen a few
days later and barely managed to get the news to Bohr
as the latter was boarding the Swedish-American liner
MS Drottningholm for New York. The soft-spoken
Bohr instantly slapped his head and said, "Oh, what
fools we have been!" In describing the process, Frisch
coined the word fission, by analogy with cell division
in biology.

It remained for three groups of physicists, including
Leo Szilard at Columbia and Walter Zinn, to demonstrate in March 1939 that neutron fission of a uranium
nucleus shakes loose several new neutrons. This
proved that chain reactions were possible, as Fermi
had conjectured. It remained for Bohr at Princeton to
calculate that only a rare form of uranium called
U-235, making up about 1 percent of the element in
nature, could sustain a chain reaction. That was why
the world hadn't already blown up on its own. To
build a chain reactor, U-235 had to be culled and

concentrated. It could be done. It could be done by the Germans. On August 2, 1939, Albert Einstein sent a letter to President Roosevelt: "Sir: Some recent work by E. Fermi and L. Szilard . . . leads me to expect that the element uranium may be turned into a new and important source of energy in the immediate future . . . and it is conceivable . . . that extremely powerful bombs of a new type may thus be constructed. . . ."

Powerful, yes. Fissioning a gram of uranium will produce about 10 million times the energy as burning a gram of coal and air or detonating a gram of TNT. Why is nuclear energy so much more potent than any form of energy known before? TNT explosions and coal-burning release chemical energy, which has been harnessed by people in one form or another for thousands of years. Chemical energy derives from rearranging the electrons in the outer parts of atoms. Nuclear energy, of the kind we've been discussing, derives from rearranging the protons in the nucleus of the atom. Because protons are confined to a much smaller volume than electrons, their electrical "springs" are much more compressed and thus much more violent upon release. Roughly speaking, nuclear energy is more powerful than chemical energy by the same factor as the atom is larger than its nucleus. (An even more powerful form of nuclear energy works by fusing small nuclei rather than fissioning large ones.)

As Soddy predicted, nuclear energy has indeed been used for peaceful purposes. The first atomic power generating plant began operation in Lemont, Illinois, in 1956. Unfortunately, nuclear power, which initially promised to be "too cheap to meter," has not yet proven its mettle economically. In 1984, the eighty-two nuclear plants licensed in the United States supplied only about 13 percent of our total electric power needs and suffered from problems with management and design. Some countries in Europe have done better, but coal and oil are still the principal workhorses of the twentieth century.

What nuclear energy has dramatically changed is the meaning of war. Each new weapon in its time seemed a giant advance over its predecessors—the Roman catapult, the medieval English longbow, gunpowder artillery in the fourteenth century, TNT in 1890—but these strides were Lilliputian by comparison to the leap from chemical to nuclear weapons. Ninety-seven out of 101 of the V-1 buzz bombs aimed at London on August 28, 1944, were intercepted—a remarkable success in defense. Had these been nuclear bombs, the four that landed, in fact just one of the four, could have annihilated the whole of the city. The United States and the Soviet Union today each possess twenty thousand such bombs, which can be launched on short notice. In our nuclear age, those ancient words of war, *defense* and *victory,* have suddenly lost their meaning. Nuclear weapons demand that we find new concepts for war and peace and weapons themselves.

Even in peacetime, nuclear weapons have violated our sense of security. In a recent national survey of high school students, done by Educators for Social Responsibility, 80 percent thought there would be a nuclear war in the next twenty years, and 90 percent of these felt the world would not survive it. How does one measure the psychological effects of these visions?

There is of late a widespread perception that technology, and nuclear technology in particular, has gained a momentum of its own and is hurling the world toward destruction. According to this belief, we humans are mere bystanders, helplessly awaiting our fate. I believe that our apparent helplessness regarding nuclear weapons originates from the *abstractness* of the danger more than our inability to stop it. After the destruction of Pompeii in A.D. 79, Mount Vesuvius exploded nine more times before another major eruption in 1631 destroyed many villages on its slopes and killed three thousand people. For six months prior, earthquakes shook the villages. Why did people continue to go about their business next to a working volcano? About seven hundred people were killed in the great San Francisco earthquake of 1906, and experts expect the area is due for another big one. Why do people continue to build their houses on the San Andreas fault? In these examples, as in nuclear war, the disaster has an all-or-nothing character, and its likelihood seems either small or incalculable. Of course, we cannot simply remove ourselves from nuclear weapons as we can from volcanoes and geological faults, but the psychology may be the same. Evidently, even with a choice to do otherwise, people will live in a dangerous situation, as long as the danger can be abstracted away.

The discovery of nuclear fission has gotten the world profoundly stuck, to use Freeman Dyson's word. Stuck in a buildup of nuclear weapons, stuck in outdated concepts of war and peace, stuck in human nature. If we can get ourselves unstuck, a thousand years from now people may well remember this era not so much for opening up the atom as for opening up ourselves.

A MODEST PROPOSAL

There are so many of my generation who have never felt a war. By and large, this is a good thing, of course, but as we postwar babies slowly climb into the seats of political power, I wonder about the consequences of today's terrible weapons coming into the hands of such innocents. Current leaders, whatever their politics, at least can recall the appalling death scenes at Hiroshima, Nagasaki, Hamburg, Tokyo, and Dresden. I've seen only photographs. I've read books. I was riveted to the television set by *The Day After* and talked it over in horror with my wife. Then, after several days, it wore off, like the memory of a nightmare.

Some of the scientists of my age, not waiting for any particular seniority, have already put their minds to designing the new generation of space-based weapons, spurred on by President Reagan's Star Wars speech in 1983. Most of these fellows haven't even seen a nuclear explosion. Since the 1963 Test Ban Treaty, there haven't been any in this country, above ground.

What we seem to be concocting, in these vastly

improved weapons now planned, is an increasingly volatile mixture of the concrete with the abstract: the weapons themselves, bristling with multiple warheads and computer chips and calculated accuracy, are there all right, occupying volume—but not of this world. Earthbound ICBMs, waiting silent and pre-programmed in their Midwest silos, are dreamlike enough. Weapons orbiting in space dissolve almost completely into a mist of make-believe.

Last Sunday afternoon, my wife was out gardening and my daughter upstairs napping. Antsy from one of the more cerebral new books on nuclear weapons, I got up from my chair, paced around the house, and hit upon a plan of action, which I hope will not be liable to the least objection. All in all, it's a modest proposal. A small country, safely distant from the superpowers, should be destroyed with nuclear weapons as the world looks on. This can be done periodically, say every twenty years, so that the carnage of nuclear destruction stays fresh in our minds. Evidently, the hundred thousand or so people killed in Hiroshima and Nagasaki were not nearly enough, so I recommend selection of a sacrificial country of about 10 million souls. As an added bonus, we could choose an economically troubled nation, now straining the world's monetary balance.

Not wanting my proposal to be taken as half-baked, I have given some attention to the necessary figures. I reckon that 10 million people, living mostly in cities, could be killed handily with no more than thirty warheads, at three hundred kilotons of TNT per warhead. (This is roughly the yield of the W78 nuclear warhead, a favorite of the United States Minuteman force and about twenty-five times more powerful than the bomb dropped on Hiroshima.) As Soviet bombs typically have a somewhat higher yield, a lesser number of them could accomplish the same work.

No sensible individual can fail to see the economy of this proposal. The total population of the world is over 4 and a half billion persons. My plan requests the sacrifice of about one quarter of 1 percent of this number. The total world supply of nuclear warheads is about fifty thousand. My plan requires only one sixteenth of 1 percent.

An essential part of the scheme, of course, is that the destruction be well documented and attended by the appropriate parties. In the last use of nuclear bombs, forty years ago, we simply did not have the interest or the equipment for adequate coverage. To this end, I would recommend, as a minimum, that the leaders of the U.S. and the U.S.S.R. and the nuclear weapons scientists concerned be bussed in to personally inspect the damage, as soon as the radiation level has slipped to a tolerable level. As to recording the event live, a very knowledgeable acquaintance of mine assures me that the United States' Big Bird satellites have the impressive ability to pick out details as small as one foot in size from an orbit of a few hundred miles above the earth. At that height the satellites would be out of harm's way and yet could photograph single charred bodies in the rubble below. The men, women, and children on the outskirts of each of the bomb blasts, not immediately killed by the explosion, could be photographed in time sequence as they develop burns, vomit with radiation sickness, and writhe on the ground for several hours or days. Shortly thereafter, the satellites would relay these pictures to newspapers, magazines, and television studios around the world. We are well accustomed in our society to viewing such mass-media spectacles, and this one should have greater redeeming social value than most.

The strong points of my plan, I humbly believe, are obvious and many. First and foremost, it would provide every person in the world with a personal and graphic understanding of the outcome of nuclear weapons. If people witnessed a sample of the destruction, then more restraint might be brought to the continuing arms race. Second, the citizens of the sacrificed countries would themselves benefit. Their economic difficulties would instantly be solved, and they could go to their deaths proudly, knowing that in their dying they were doing the rest of the world a good turn. Third, the idea is thrifty, requiring only a small portion of the world's population and nuclear resources. Fourth, the scheme should easily receive the cooperation of the superpowers, who would have little to lose and much to gain. And finally, owing to the efficiency of the weapons involved, the entire business excepting the publicity could be finished in a tidy few days.

We live in a new Age of Reason. It is time we acted accordingly. What could be more reasonable than the expense of a few million lives, to gain a practical education in nuclear weapons?

LOST IN
SPACE

My first face-to-face meeting with scientists working on space weapons happened on a visit to the Hudson Institute in January 1979. Jimmy Carter was still in the White House and Ronald Reagan's Star Wars speech was more than four years down the road.

After strolling across the serene and spacious grounds of the institute, which is nestled in a bucolic setting overlooking the Hudson River, I was led into the cozy office of a physicist who could scarcely wait to tell me about new developments in "particle beam" weapons, to be stationed aboard orbiting satellites in space. After a vivid description of intense beams piercing earthward from above, he excitedly showed me some artists' sketches of what the things might look like and, as I recall, even had a papier-mâché model, which he let me hold briefly. For a moment I got caught up in the air of derring-do, and wondered what Santa might bring *me* next year.

That childlike and visceral attraction to the new generation of nuclear weapons in space is real, dangerous, and seldom discussed. The technical issues, on the

other hand, have received a lot of attention. To sum-
marize briefly, there is a growing consensus among
most weapons experts that a space-based ballistic missile defense system is probably not workable for another twenty years and, even if eventually deployed,
would be costly and destabilizing. In the words of
Hans Bethe, the needed technologies are "far beyond
the state of the art."

While countermeasures, like decoys and hardened
boosters, appear abundant and relatively cheap, the
price of a working system would be staggering. To aim
a defensive laser beam at the soft parts of a missile a
thousand miles away requires the angular resolution of
the space telescope, which runs to a billion dollars.
Several hundred ballistic missile defense satellites
would be needed to give adequate coverage of the
Soviet Union, adding up to at least several hundred
billion dollars, not to mention the incalculable cost of
a new arms race likely to follow.

Once operational, an effective space-based system,
in unison with offensive weapons, could give the de-
ployer a first-strike capability, inviting pre-emptive attack by the other side. The high vulnerability of such
a system further encourages attack. And any orbiting
weapon in space, offensive or defensive, threatens the
surveillance and communications satellites used since
the early 1960s and considered vital for national security. All of the above arguments apply whether the
ballistic missile defense system is nuclear or not. A
nuclear space-based ballistic missile defense system
would, in addition, violate the 1967 Outer Space
Treaty and the 1972 Anti-Ballistic Missile Treaty.
When it comes to nuclear weapons, all agreements
between us and the Soviets seem especially precious.
These are a few of the technical worries. Apparently
they have had some effect among Congressmen, but

President Reagan shows no signs of reconsidering his proposal for a space-based defensive system.

Technical issues aside, however, the glamour of Star Wars still shimmers and beckons. Millions of us, children and grownups alike, saw the movie and were mesmerized by images of death-dealing laser rays, sleek aircraft shooting it out in space, and handsome young men battling the forces of evil. These heady visions seep into the unconscious and resonate with the leftover daydreams of little boys. The space age is here at last, and no one—teacher or businessman or senator—wants to be left behind. "Seize the high ground before the Russians do" is a familiar bugle call from the Air Force, which established its Space Command in 1982.

Scientists are needed to work on these things and scientists, as C. P. Snow has reminded us, are not much different from other people. The team of fighter jocks immortalized in Tom Wolfe's *The Right Stuff* seems to have been curiously reincarnated in the dozen or so young physicists "pushing back the edge" in space weapons design in O Group at Lawrence Livermore National Laboratory.

Profiled in William Broad's recent book *The Star Warriors,* these intellectual test pilots are mostly in their twenties, and are all male. They inhabit a world of empty Coke bottles and all-night bouts with top-secret research, and they share an admiring respect for each other's brain power. A "right stuff" ethic flourishes in all areas of science. But in O Group this is combined with the glamour of space, the thrill of inventing new kinds of nuclear weapons, and youthful idealism.

Says O Group physicist Lawrence West, age twenty-eight, "We can try to negotiate treaties and things like that. But one thing I can do personally, without having to wait for arms control, is to develop the technology to eliminate them myself, to eliminate offensive nuclear weapons." What more dangerous creature than the inexperienced macho, armed here with pencil and paper? Chuck Yeager, Gordon Cooper, and John Glenn all prided themselves on hanging their hides over the edge. How much hide hanging has been done and can be done by the fellows of O Group? None has seen a nuclear explosion. Even the few old men of my generation, in their midthirties, were born after World War II, after Hiroshima and Nagasaki, and learned about the Manhattan Project from books.

A related motive to watch closely is the love of technology for its own sake: from greeting cards that sing Happy Birthday when opened, to F-15s that turn corners at greater acceleration than pilots can endure. In theoretical physicists, this translates to pursuit of intellectually interesting problems—wherever they lead. As West says proudly, "The number of new weapon designs is limited only by one's creativity." Compare Robert Oppenheimer's comment thirty years ago: "When you see something that is technically sweet you go ahead and do it and you argue about what to do about it only after you have had your technical sweetness."

It is difficult to find fault with the argument that basic research in space weapons should continue. We had better at least find out what the laws of physics allow. Most likely the Soviets will. And it is conceivable, as the Harvard Nuclear Study Group and Freeman Dyson have suggested, that our security in the long term might best be served by replacing our current nuclear strategy of mutually assured destruction with one of defense, possibly from space. Research should continue—but soberly, with both feet on the ground.

Sirens of the unconscious call us to Star Wars: glamour, novelty, childhood fantasies, macho power, technical narcissism. It would seem wise to bring these psychological motives into daylight, to attach as much importance to them as to the technical issues. The weapons themselves are unthinking, but their creation and deployment spring from the human mind.

THE DARK NIGHT SKY

OSTENTATIO: Tell me, Timidio, why does it get dark at night?

TIMIDIO: Do you take me for a fool? It gets dark because the sun sets.

OSTENTATIO: But what about the remaining luminous stars? Why does their accumulated light not keep the heavens bright?

TIMIDIO: Because their great distance makes them too dim to be of any consequence. You are wasting my time, Ostentatio.

OSTENTATIO: But you agree there may well be an *infinite* number of stars, scattered throughout an infinite universe, each adding its light to the sky?

TIMIDIO: Yes. But they appear increasingly faint with distance.

OSTENTATIO: Then the question hinges, does it not, upon a contest between the increasing numbers of more distant stars and the decreasing amount of light from each of them. Now I will show you by pure reason that the former effect dominates, so that the sky should be constantly ablaze with light.

First, it is easily established that the apparent brightness of any source of light is reduced by four for every doubling of its distance away. To show this, imagine placing a candle at the center of a spherical balloon, which will glow as the candlelight strikes its surface. Next double the radius of the balloon, quadrupling its surface area. Since the same amount of light must now illuminate four times the surface area, each square inch of surface, now twice the distance from the candle, receives only one-quarter the light it did before.

With this result in hand, consider next the accumulated light from the stars within 100 trillion miles of earth, then within 200 trillion miles of earth, then 400 trillion miles, and so on, assuming the stars are evenly spaced. With every doubling of the distance, the volume of space, and hence the number of stars included, increases by eight. But the received light from each star, on average twice as far from us, decreases by only four. Thus the *combined* starlight received on earth doubles with each doubling of the radius of space considered. Continue this indefinitely, to infinite distance, and the accumulated light from stars must mount indefinitely.

Again, I ask you, why does it get dark at night?

TIMIDIO: I am baffled.

The above paradox, in different form, was first publicly stated by astronomer Edmund Halley. Halley read his disturbing conclusion, and its supporting arguments, before the Royal Society in London in 1721. Isaac Newton, chairing the session, had somehow overlooked the puzzle entirely. But the geometrical logic was inescapable. Later restated in other countries, where it was also observed to get dark when the sun went down, this conundrum eventually became known as Olbers' paradox, after its reassertion by astronomer Heinrich Olbers in 1826. It was correctly resolved only recently.

Critical to posing Olbers' paradox is, of course, the assumption of a spatially infinite universe. Although stretching the imagination, this was accepted in Halley's day and is still considered viable by modern cosmologists. Partial credit for originating the idea goes to Englishman Thomas Digges, who wrote of it in 1576. He was the first Copernican to replace the concept of a spatially finite universe, inherited from Aristotle, with that of an infinite one. Perhaps the clearest and most influential advocate of an infinitely extended cosmology was Giordano Bruno, an Italian scholar who, for this and other indiscretions, was burned at the stake in 1600. Bruno also recognized something we now take for granted: that sense perception alone cannot form the basis for scientific knowledge. In his *On the Infinite Universe and the Worlds,* written in 1584, he writes, "No corporeal sense can perceive the infinite. . . . Since we have experience that sense-perception deceiveth us concerning the surface of this globe on which we live, much more should we hold suspect the impression it gives us of a limit to the starry sphere." Indeed, since the seventeenth century, telescopes have peered out farther and farther at stars and galaxies unseen by the unaided eye.

All of which brings us back to Olbers' paradox. Two centuries of proposed resolutions—ranging from a suggested failure of infinitesimal increments of light to accumulate, to an absorbing medium in space—have not held their ground. Much of the modern explanation of why it gets dark is that the universe is only 10 or 15 billion years old. Since light travels at a finite speed, only the light from a distance of 10 or 15 billion light years has had time to reach our planet. That is, even if space extends infinitely far, we can glimpse only a portion of that space. For even the largest of telescopes, there is an edge to the visible universe. Each day that edge retreats outward a bit more. And within the visible universe, the accumulated light from all stars and galaxies is not enough to spoil the darkness of the night.

Incredibly, no one happened upon this argument until the 1920s. At that time, observational evidence for the outward motion of distant galaxies forced astronomers to surrender their mistaken belief in a changeless universe, without beginning. This cherished belief, like that in a spatially finite universe, was another gift from Aristotle. A hidden assumption in Olbers' paradox is that all the stars in the universe have had time to send their light from there to here —guaranteed only if the universe has existed forever. With some plucky thinking, including the disposition to throw over Aristotle, Olbers' paradox could have been put to rest long ago. The necessary ingredients have been around. That the speed of light is finite has been known since the work of the Danish astronomer Ole Roemer in 1676. (Roemer, in fact, met Halley and Newton on a visit to England in 1679.) That the universe might be impermanent rather than permanent seems no less compelling a possibility than that space might be infinite rather than finite. Nearly eight hundred years ago, Moses Maimonides locked horns with Aristotle point for point on whether the universe had a beginning. Still, it is always easier to argue with hindsight and exceedingly difficult to re-create cultural and intellectual barriers once broken.

In my view, what is fascinating here is not that Olbers' paradox *should* have been unraveled much earlier, but that it *could* have been, with the naked mind. Clearly, telescopes can see what eyes cannot, and ultrasound detectors can hear what ears cannot. Far more surprising, and exquisite, is that pure thought can probe the extrasensory universe. Why this works —the mathematics and the logic, the human way of thinking—is a mystery, but it keeps us in business. Thus did James Maxwell correctly predict electromagnetic waves in 1864, Arthur Eddington correctly propose nuclear reactions for the power source of stars in 1920, and Paul Dirac correctly predict antimatter in 1928. And with pure thought, plus the observation that it gets dark at night, we can reason backward and deduce that our universe, extending countless billions of light years into space, has not been forever.

ELAPSED
EXPECTATIONS

The limber years for scientists, as for athletes, generally come at a young age. Isaac Newton was in his early twenties when he discovered the law of gravity, Albert Einstein was twenty-six when he formulated special relativity, and James Clerk Maxwell had polished off electromagnetic theory and retired to the country by thirty-five. When I hit thirty-five myself, I went through the unpleasant but irresistible exercise of summing up my career in physics. By this age, or another few years, the most creative achievements are finished and visible. You've either got the stuff and used it or you haven't.

In my own case, as with the majority of my colleagues, I concluded that my work was respectable but not brilliant. Very well. Unfortunately, I now have to decide what to do with the rest of my life. My thirty-five-year-old friends who are attorneys and physicians and businessmen are still climbing toward their peaks, perhaps fifteen years up the road, and are blissfully uncertain of how high they'll reach. It is an awful thing, at such an age, to fully grasp one's limitations.

Why do scientists peak sooner than most other professionals? No one knows for sure. I suspect it has something to do with the single focus and detachment of the subject. A handiness for visualizing in six dimensions or for abstracting the motion of a pendulum favors a nimble mind but apparently has little to do with anything else. In contrast, the arts and humanities require experience with life, experience that accumulates and deepens with age. In science, you're ultimately trying to connect with the clean logic of mathematics and the physical world; in the humanities, with people. Even within science itself, a telling trend is evident. Progressing from the more pure and self-contained of sciences to the less tidy, the seminal contributions spring forth later and later in life. The average age of election to England's Royal Society is lowest in mathematics. In physics, the average age at which Nobel Prize winners do their prize-winning work is thirty-six; in chemistry it is thirty-nine, and so on.

Another factor is the enormous pressure to take on administrative and advisory tasks, descending on you in your mid-thirties and leaving time for little else. Such pressures also occur in other professions, of course, but it seems to me they arrive sooner in a discipline where talent flowers in relative youth. Although the politics of science demands its own brand of talent, the ultimate source of approval—and invitation to supervise—is your personal contribution to the subject itself. As in so many other professions, the administrative and political plums conferred in recognition of past achievements can suffocate future ones. These plums may be politely refused, but perhaps the temptation to accept beckons more strongly when you're not constantly galloping off into new research.

Some of my colleagues brood as I do over this passage, many are oblivious to it, and many sail happily ahead into administration and teaching, without looking back. Service on national advisory panels, for example, benefits the professional community and nation at large, allowing senior scientists to share with society their technical knowledge. Writing textbooks can be satisfying and provides the soil that allows new ideas to take root. Most people also try to keep their hands in research, in some form or another. A favorite way is to gradually surround oneself with a large group of disciples, nourishing the imaginative youngsters with wisdom and perhaps enjoying the authority. Scientists with charisma and leadership contribute a great deal in this manner. Another, more subtle tactic is to hold on to the reins, singlehandedly, but find thinner and thinner horses to ride. (This can easily be done by narrowing one's field in order to remain "the world's expert.") Or simply plow ahead with research as in earlier years, aware or not that the light has dimmed. The 1 percent of scientists who have truly illuminated their subject can continue in this manner, to good effect, well beyond their prime.

For me, none of these activities offers an agreeable way out. I hold no illusions about my own achievements in science, but I've had my moments, and I know what it feels like to unravel a mystery no one has understood before, sitting alone at my desk with only pencil and paper and wondering how it happened. That magic cannot be replaced. When I directed an astrophysics conference last summer and realized that most of the exciting research was being reported by ambitious young people in their mid-twenties, waving their calculations and ideas in the air and scarcely slowing down to acknowledge their predecessors, I would have instantly traded my position for theirs. It is the creative element of my profession, not the exposition or administration, that sets me on fire. In this regard, I side with the great mathematician G. H. Hardy, who wrote (at age sixty-three) that "the function of a mathematician is to do something, to prove new theorems, to add to mathematics, and not to talk about what he or other mathematicians have done."

In childhood, I used to lie in bed at night and fantasize about different things I might do with my life, whether I would be this or that, and what was so delicious was the limitless potential, the years shimmering ahead in unpredictability. It is the loss of that I grieve. In a way, I have gotten an unwanted glimpse of my mortality. The private discoveries of new territory are not as frequent now. Knowing this, I might make myself useful in other ways. But another thirty-five years of supervising students, serving on committees, reviewing others' work, is somehow too social. Inevitably, we must all reach our personal limits in whatever professions we choose. In science, this happens at an unreasonably young age, with a lot of life remaining. Some of my older colleagues, having passed through this soul-searching period themselves, tell me I'll get over it in time. I wonder how. None of my fragile childhood dreams, my parents' ambitious encouragement, my education at all the best schools, prepared me for this early seniority, this stiffening at thirty-five.

LET THERE BE LIGHT

Not long ago, I had a conversation with Margaret Geller, of the Harvard-Smithsonian Center for Astrophysics in Cambridge. We were talking about light. Dr. Geller, an astronomer in her late thirties, spends her time collecting and analyzing the light from distant galaxies. A couple of months earlier, she and two of her colleagues had astounded the scientific world with evidence that galaxies seem to congregate on the surfaces of bubblelike structures, each about a thousand times the diameter of one galaxy. This could be the big picture, the way the universe looks to God. And it is all inference, based on light. Light is the only physical reality we've gotten from other galaxies. "When I'm working at a large telescope," said Dr. Geller, "it is eerie to realize that the light coming in has been traveling through space for hundreds of millions of years, undisturbed, and just at that moment is stopped by my telescope." Much of the light stopped by large telescopes these days is never seen by a human eye. It gets recorded by electronic devices, and digitized, and stored on computers.

. . .

One day in 1655, a Jesuit priest named Francesco Maria Grimaldi did a simple experiment with light that had never been done before. He let a shaft of sunlight enter through a tiny hole into a darkened room, travel past a small object in its path, and fall upon a white screen. The object's silhouette on the screen was odd. Where there should have been only black shadow—directly behind the object—there was an invasion of light, and where there should have been only light, there were curious dark marks. Evidently, the light rays in Grimaldi's room in Bologna were not traveling in straight lines, as predicted by scientists and philosophers through the centuries. But Grimaldi, like his countryman Galileo, paid no attention to previous authorities and continued his experiments for the remaining eight years of his life. Grimaldi had discovered that light was similar to a fluid. Upon meeting an obstacle, this fluid will split into secondary ripples, which can then overlap and interfere with each other downstream, producing a complex pattern of light and dark. Grimaldi had discovered that light was a wave.

Almost three and a half centuries later, Philipp Lenard, a physicist and the son of a wine maker, discovered something quite different about light. Lenard irradiated some metal with the intense light from a carbon arc lamp, knocking electrons in the metal out of their atoms, and found that the energy gained by each electron didn't depend on the intensity of light. A greater intensity of light set free more electrons, but it didn't increase the energy of each electron released. This was puzzling. If light was a fluid, a stronger flow of the fluid should have imparted more energy to each electron. Einstein proposed an explanation a few years later, in 1905. Light flows not in a continuous stream, but in a series of individual particles, called photons, each carrying the same energy. Increasing the intensity of light increases the number of photons, but not the energy of individual photons. According to Einstein's analysis, each electron in Lenard's experiment always absorbed just one photon, and therefore its energy gain did not depend on the intensity of the light. Lenard had discovered that light was a particle.

Somehow, in a way that physicists can calculate and can measure but still can't picture, light is both a particle and a wave. Light is also pure energy—a union of electric and magnetic force fields that have broken free of their generators and magnets and taken on a life of their own, wiggling back and forth and flying at great speed through substance or vacuum—and none of my colleagues can picture this either. It is far easier to visualize sound. Sound is mechanical. Sound is a jostling of molecules. A person yells at one end of a room and her voice compresses a bunch of air molecules, and this increases the pressure, which pushes on a neighboring bunch of molecules, and so on, and this compression and pushing of molecules works its way across the room until it strikes someone's ear. We can't directly see the molecules, but we can imagine them, because we can see marbles and then mentally shrink the marbles to molecules. Light, however, doesn't need molecules, or anything material. The physicist Richard Feynman, who is not only a Nobel laureate but also a genius at explaining abstruse phenomena with homespun analogies, claims that he can picture invisible angels but not light waves.

Through history, our understanding of light has reflected our science as a whole. As our knowledge of light has become more quantified but less intuitive, so has our knowledge of all nature.

In earlier times, light was much simpler. In ancient Greece, and long afterward, two competing theories of light and vision held sway: the visual ray theory of the Pythagoreans and the eidola theory of the Atomists. According to the Pythagorean school, which included such luminaries as Plato and Euclid, the light we see by is a divine fire, originating inside the body. The eye is like a flashlight. It sees the outside world by emitting light rays, which travel to objects, illuminate them, and reflect back into the eye. (Even today, we have idioms that hark back to this theory; for example: "He sent her a piercing glance.") Consider the first postulate in Euclid's *Optics:* "The rays emitted by the eye travel in a straight line."

In contrast to the Pythagoreans, the Atomists believed that the light used in vision originates outside the eye. All objects, in this theory, are continuously throwing off delicate images of themselves, called eidola, which travel to the eye, enter the pupil, and allow sight. Furthermore, the eidola are material. Although fragile, they have weight. They are made of atoms. The Atomists, who included Democritus and Epicurus and later the Roman poet Lucretius, constructed the entire universe from atoms. All was material. Lucretius (c. 99–55 B.C.), in his great didactic poem *De Rerum Natura,* says that sight, like hearing and smell, is agreeable when the atoms carrying it are smooth and round, and disagreeable when the atoms are crooked and hooked. Light passes through a lantern made of animal horn, but rain doesn't, because particles of light are smaller than those of water. In the middle of his poem, Lucretius makes a wise observation that eventually conquered the visual ray theory:

Bright objects, moreover, the eyes avoid and try not to see. The sun actually blinds if you persist in staring against it, because its own power is great, and from on high through pure air the images come heavily rushing, and strike the eyes so as to disturb their structure.

If the eye sees by sending out its own rays of light, there is no reason why looking at one object would bring pain while looking at another would not. This argument was repeated by the Arab physicist Ibn al-Haytham (965–1039), when it finally stuck, and the visual ray theory was not taken seriously again. The eidola theory of course had its own problems. No one ever explained exactly how the fragile eidola could squeeze through the pupil intact.

In about the year 1800, scientists realized that some kinds of light couldn't even be seen by the eye. Invisible light was first discovered by the astronomer William Herschel. Herschel is better known for finding the planet Uranus, but he also did some remarkable experiments with the colors in sunlight. Since the time of Grimaldi and Newton, it had been known that white light could be split into colors by a prism. Grimaldi correctly identified these different colors with different wavelengths of light; that is, with waves of differing distance between successive crests. (For example, the wavelength of violet is about forty millionths of a centimeter; that of red is about seventy millionths of a centimeter.) Ordinary light contains a mixture of different wavelengths. A prism changes the direction of each incoming wave by an amount varying with its wavelength, and that is why the different colors fan out after leaving the prism.

Herschel wanted to measure the individual intensity of each color in sunlight. (Each source of light has a unique mixture of colors.) To do this, he passed sunlight through a prism and projected the outgoing strip of colors onto a row of thermometers. Each thermometer would heat up in proportion to the intensity of light falling on it. After Herschel had lined up his thermometers to sample all of the visible spectrum, beginning with red at the left and ending with violet on the right, he added some thermometers on the left. And to his surprise, he found that they heated up too. Evidently, there were invisible colors redder than red, the so-called infrareds. Other scientists soon ventured beyond the other end of the visible spectrum and discovered the ultraviolets. The ultraviolets are the dog whistles of color. By the end of the nineteenth century, scientists knew that light comes in a vast range of wavelengths, only a small portion of which is visible. Radio waves are light waves of wavelengths much longer than those visible. X-rays are light waves of very short wavelengths. From X-rays to radio waves is more than a factor of a hundred million in wavelength. From violet to red is less than a factor of two.

I remember when I first learned that light was a wave of electric and magnetic force fields. It was in freshman physics. For months, we had been gingerly stepping through classical physics, beginning with mechanics and gravity and then inching our way through electricity and magnetism. We had been told that a charge of electricity attracted or repelled other charges through an invisible "electric force field" around it. Likewise, a magnet projected an invisible "magnetic force field." These mysterious force fields gained some respectability when we sprinkled iron filings near a magnet and saw them twist and arrange themselves in a beautiful pattern, clearly in the grip of something, even though we couldn't see what it was. This was all fun, but it had nothing to do with light. Then one day, I came to a chapter in the book where the equations for electricity and magnetism were combined in a certain way, which I barely could grasp with my math at the time, and out leaped oscillating waves of electric and magnetic force fields, detached from their charges and magnets and traveling on their own. Once off and running, the force fields re-created and sustained each other. The speed of these "electromagnetic waves" could be computed from electric and magnetic parameters we had measured in the lab. It happened to be 186,000 miles per second, exactly the speed of light. Electromagnetic waves and light waves were one and the same. I was astonished that cats' fur and magnets could tell me about light.

The person who discovered the connection between light and electromagnetic waves was James Clerk Maxwell. Einstein, who was born the year Maxwell died, once said that the change in our concept of nature due to Maxwell was "the most profound and the most fruitful that physics has experienced since the time of Newton." The unleashed force fields of Maxwell started the revolt that overthrew the purely mechanical world of Newton.

Maxwell, a Scottish descendant of the Clerks of Penicuik and the Maxwells of Middlebie, was a brilliant theoretician. After graduating second wrangler and first Smith's prizeman from Cambridge University in 1854, he began his investigations of electricity. Since the 1820s, it had been known from the experiments of Hans Oersted, Michael Faraday, and André Ampère that electricity and magnetism were related. A magnetic force field could create an electrical current, and vice versa. Equations had been developed to describe these relationships. But Maxwell noticed the equations were not consistent. Something needed to be added to one of them. In 1864, using pencil and paper, Maxwell found the missing piece. The augmented equations, forever after held sacred by physicists and known as Maxwell's equations, predicted the existence of electromagnetic waves and required that these waves travel with a particular speed. That speed was the same number as the measured speed of light. To appreciate this discovery, one must realize that nothing about light had been used in the derivation of Maxwell's hypothetical waves. "The agreement of these results [the two speeds]," Maxwell wrote, "seems to show . . . that light is an electromagnetic disturbance propagated through the field according to electromagnetic laws."

Maxwell's equations also show that all electromagnetic waves are first created by the rapid motions of electrically charged particles. Likewise, after traveling across a room or between stars, electromagnetic waves are detected by setting in motion the charged particles in the receiver. The vibration of atoms in a distant star creates light, which travels through space and vibrates the electrons in our eye—and that is why we see the stars.

Maxwell was lucky that the speed of light was known in his day. I sometimes wonder how much he would have been held back if this speed were too great to be measured. In fact, it is so enormous that people as recent as Kepler and Descartes believed it was infinite. The speed of light was first measured in 1676, by Ole Roemer, a Danish astronomer working in Paris. Roemer used an astronomical method. Since the invention of the telescope, around 1600, astronomers had been watching the moons of Jupiter as they orbited about the planet. Each moon disappears when it moves to the far side of Jupiter and reappears when it emerges on the near side. These measurements were of practical use to mariners, since the regular and frequent occultations of Jupiter's moons could be used as celestial clocks, which could then be compared to the local position of the sun and stars to establish longitude and latitude. According to the laws of gravity, the same amount of time should pass between each occultation of a particular moon. However, it was noted by Roemer and a few others that the occultations were not quite regular: they happened more often when the earth was approaching Jupiter and less often when it was moving away. Roemer correctly interpreted this effect as arising from the finite speed of light. The light rays announcing the start of a new occultation would take less time to arrive at the earth if the earth had moved closer to Jupiter since the previous occultation, and more time if the earth had moved farther away. In September 1676, Roemer announced to the Académie des Sciences that the November 9 occultation of Jupiter's inner moon Io would happen ten minutes behind schedule. He was right. In the process, he had determined for the first time in history the speed of light, accurate to about 30 percent.

In 1878, the year before he died, Maxwell wrote an article titled "Ether" for the ninth edition of the *Encylopaedia Britannica*. The ether was an hypothesized substance that filled all of space and was responsible for transmitting electromagnetic waves. Just as sound waves were carried along by the vibrations of molecules, each bumping the next, electromagnetic waves were supposedly carried along by the mechanical vibrations of this very flimsy stuff called ether. It had to be flimsy because it had never been measured. Nevertheless, people firmly believed that light, like every other wave motion they knew about, needed a material medium for propagating it. The ether was part of the mechanical world view handed down by Newton. It was an old idea. Maxwell accepted it. In his article, he wrote: "There can be no doubt that the the interplanetary and interstellar spaces are not empty but are occupied by a material substance or body, which is certainly the largest, and probably the most uniform, body of which we have any knowledge."

In 1887, Albert A. Michelson, a physicist at the Case School of Applied Science in Cleveland, and Edward Williams Morley, a chemist from next-door Western Reserve University, performed a landmark experiment to measure the ether. More precisely, Michelson and Morley attempted to measure the velocity of the earth relative to the ether. The idea behind their experiment is simple. If light is a wave traveling in an ether, and if the ether is moving, then the net speed of a light ray should change as it changes direction—just as the speed of a boat in a moving river, as measured by someone on shore, changes when the boat first goes upstream and then downstream. As the earth moves through space, the hypothesized ether flows past it, like a current. By measuring the speed of light traveling in different directions relative to this current, the velocity of the current can be found (or, equivalently, the velocity of the earth through the ether). This is what Michelson and Morley did. To their disappointment, they measured the same speed for light traveling in any direction. They had failed to find any evidence of the ether.

But physicists clung to the ether and invented all sorts of theories to save it. Finally, after some thirty years of complaining and excuses, most people acknowledged that the ether did not exist. The electromagnetic force fields of light didn't need an ether. They could travel perfectly well in a vacuum. Light was not the mechanical vibration that Maxwell believed in, but a vibration of energy. And the purely mechanical world was no more. It is ironic that the man who most understood light in the last century, and indeed gave us the equations for light that we use today, had an incorrect physical picture in his head. If that picture had been right, Richard Feynman could probably imagine light waves today.

The demise of the ether was even more discomforting in other ways. It eventually demanded a new concept of time. The absence of an ether, and the uniform speed of light in any direction, implied that it was not possible to measure one's motion by looking at light rays. In 1905, the twenty-six-year-old Albert Einstein calmly postulated that any observer, *regardless of his motion,* would measure the same speed for a passing light ray. Consider what this means. A light ray traveling at 186,000 miles per second inside a car, which itself is going at 100,000 miles per second, goes past someone on the sidewalk at 186,000 miles per second! This postulate totally violates everyday experience, but it has been proven right. To be right, it requires that the clock of the person in the car and the clock of the person on the sidewalk tick at different rates. Speeds are measured with clocks, and if speeds behave strangely, so must clocks and time itself.

Astronomers measure distances in terms of the light year, the distance light travels in a year. The nearest star is about five light years away. As I write these words, the image of Einstein, hunched over his desk in a patent office and dreaming up relativity, is about 80 light years from earth and racing outward; Maxwell, calculating force fields at his country estate in Galloway, is about 100 light years out; Grimaldi, lit with the sunlight he has just dissected, is some 330 light years out; Lucretius, pondering atoms and eidola, is some 2030 light years out. And for thousands of light years beyond are the images of earthlings who have been fascinated and mystified by light.

HOW THE CAMEL
GOT HIS HUMP

It is early evening and I am putting my daughter to bed. She sits beside me in her yellow pajamas, with her head against my shoulder. For the third time, we are making our way through the *Just So Stories*. My daughter wants to know if the magic Djinn in charge of All Deserts could really cause the Camel's back to puff up so suddenly, and what good is the hump anyway. She has asked this before. Tonight, I am prepared, having looked up camels in the library and talked to some knowledgeable friends. The hump, I explain, is made of fat, which all animals need to live on when they can't find food. The camel keeps all its fat in one place, on its back, so that the rest of its body can cool off more easily. Staying cool is important in the desert. The penguin, on the other hand, needs to stay warm, and spreads its fat in a thick layer all over its body, like a blanket. I tuck my daughter in.

"Daddy, camels are wicked smart, aren't they," she says, yawning.

"Not really," I say. "Camels didn't figure things out on their own. Nature spent millions and millions of years working on camels, making lots of mistakes until they came out right."

I turn off the lights. The streetlamp outside shines through the bedroom window. I think of my visit to New York last week, coming into the city at night on a bus, with the buildings and towers all lit up, slender and beautiful and fragile, like miniatures. And then, on the Queensboro Bridge, with the streetlamps passing one by one, the light pulses on the vinyl seat in front of me, making it look like throbbing skin, the very thin skin on a person's throat, quivering with each pulse of blood in the veins underneath.

My daughter sneezes. "Guess what I made in school today, Daddy," she says.

"What?"

"A Pilgrim, for Thanksgiving. And before that, I climbed up to the fourth rung of the ladder. The fourth rung. Mrs. Gauthier saw me."

I kiss her and walk to the window. "Come look at the moon with me," I whisper. She gets out of bed and tiptoes barefoot across the carpet. We open the white shutters.

"Men have gone to the moon and walked on it," I say. "Just a few years ago."

The night is broken by the sound of a car down the street.

I look at the moon again, hanging in space, and I imagine giant wheels of steel, rotating silently in the darkness overhead. I imagine thousands of satellites whizzing around the planet in all directions, narrowly missing each other. I imagine smooth cylinders suddenly launched upward, lighting the night with the red fire from their engines, arcing toward cities. New toys of new creatures. And below, the ancient earth waits.

"Back to bed," I whisper to my daughter. I tuck her

in again, folding the blanket carefully across her chest.

"Daddy," she says, "will you read to me again about the Djinn, and how he made the hump puff up with magic?"

"Another night," I answer.

A MODERN DAY YANKEE IN A CONNECTICUT COURT

One day last week, I found in my mailbox the following account from a man I know slightly.

If you look up the Howe family in Hartford, Connecticut, you'll find that a curious item has been passed down among the old family heirlooms. It is a ballpoint pen, found among the personal effects of one Phineas Howe, who practiced law in the last century. The pen is cracked and dirty, but it is unmistakably what it is —a Bic ballpoint. No one living except me knows how Phineas got that pen. Here is my story.

I am an assistant manager of a department store and live in the Boston area. Although I spend most of my time wrestling with inventories, I think of myself as having a decent general knowledge of the world. On the evening of August 9, 1985, I was relaxing at home after a long day at work, when I leaned over to take off my shoes. I must have struck the bookshelf, because my Panasonic home entertainment center came crashing down and hit me in the head.

When I came to, I found myself lying in a meadow, next to a dirt road. Peering down at me was a man in a buggy. He was wearing funny-looking baggy pants and suspenders. As I began to get to my feet, the man spoke to me.

"You from New York?"

"New York?" I repeated, gingerly exploring the bump on my head.

"Yep, New York. I don't know where else clothes like that come from."

"Where am I?" I asked slowly.

The man looked at me as if I were nuts. "You're in back of the Armory," he answered.

"The Armory?"

"The Colt Armory," he said. "In Hartford."

"Hartford," I shouted. "What day is it?" If I'd been out several days, I was in deep trouble with Mr. Godine, my boss.

The man in the buggy shook his head and smiled sympathetically. "It's Monday," he said. "Monday the ninth. Now why don't you just come along with me to the Armory. We've got a doctor there."

"Have you got a telephone?" I asked quickly.

"Due for one at the first of the year," he said. "Time being, we've got two good telegraph lines."

I silently climbed into the buggy. The man gave a pull on the reins, his horse gave a snort, and we trotted off down the road.

"By the way," I said, "I know this may sound stupid, but what month is it?"

"August," said my new acquaintance. And then, incidentally, "Eighteen eighty."

Pretty soon, two large smokestacks billowing smoke came into view, then a whole complex of buildings. There were three main buildings, each four stories high, connected together in the form of a capital H. Three sides of the factory were enclosed by a wide dirt road and a wooden fence. The fourth side bordered a river. Through the trees and the buildings, I could just make out the masts of what must have been some large steamboats or schooners at dock.

We rode through the main entrance of the Armory and parked our horse and buggy next to another horse and buggy. Before I had walked ten feet, a crowd of workers, all dressed in baggy pants and suspenders, gathered around and were gawking at me. Either I was crazy or they were, and they had the majority. I made the mistake of being honest. When I told them it was August 9, 1985, the last I remembered, the men roared with laughter. I told them where I worked and where I lived. I began reciting recent presidents: "Nixon, Ford, Carter, Reagan . . ."

"Mind yourself," blurted one burly guy, "or we'll have you carted off to the Retreat for the Insane."

I decided it was time to take a quiet walk around downtown Hartford, so I politely asked for directions and left. By now, I was mostly sure that somehow I had gotten bumped back in time.

It was about a mile and a half to the center of town. On the way, I passed several more horse-and-buggy combinations. I also passed a group of people cheering and hooting as if a contest were about to begin. When I got a better look, I saw that it was a race between a horse and a bicycle. A boy of about twelve, wearing a red cap and striped knickers, straddled the bike and could hardly wait to launch himself. Everyone seemed awed by the bicycle, except two or three older men who were scoffing at it.

I walked on. I have to admit that within a few minutes of wandering around town, I forgot my predicament. It was a warm summer morning and the air smelled sweet. The dirt streets were wide and easy, the traffic was light, and the stores weren't selling the usual. One firm, named Wm. H. Wiley, produced something called over-gaiters. A store called Smith

Medicated Prune Company offered free samples. Around the corner were the smokestacks of an enormous Pratt and Whitney factory, advertising machine tools, gun tools, tools for sewing machines, and steam engines, all produced with "precision, durability, and complete adaptation of means to ends." Another company flew a flag with the motto: BETTER MACHINES FOR A BETTER LIFE. Posted against the first-floor window was a drawing of "Thomas A. Edison's new talking machine," showing a long cylinder, mounted at both ends and pressed against some kind of earphone or speaker in the middle. In the picture, a cheerful woman leaned over the machine and was turning the cylinder with a crank.

Getting tired, I spotted a park bench and sat down. I was oddly excited. There was a feeling of progress in the air. Technology was booming. Life was improving.

Then it struck me how to prove who I was. I was a man of the twentieth century. I could reveal to them the wonders of modern technology. They would have to believe me. My knowledge would speak for itself. And there was something else. I'd been taking orders for years. It was about time to be the person in charge. I began feeling a heady sense of power.

The Colt Armory seemed the best place to start, since I already had acquaintances there. I ran all the way and immediately sought out Amos Plimpton, the fellow who had taken me into his buggy. He was operating a metal stamping machine when I found him.

"Mr. Plimpton," I said out of breath, "give me just fifteen minutes with your best machinists. I've got some very interesting things to tell them about the competition. I promise it will be worth their while."

Plimpton miraculously consented, probably giving in to his good Yankee business sense.

After about twenty men had congregated in Plimpton's shop, I got under way. I figured I would start easy, maybe with cars, and work up to VCRs. "Gentlemen," I began, "let me tell you about a very advanced means of transportation called an automobile. I think you've probably got the tools to build one right here in this shop." Silence. I continued. "An automobile has a gasoline engine that revs up when you put your foot on the accelerator, and it will carry you along the road at up to a hundred miles per hour." I smiled.

"How does this gasoline engine work?" asked one fellow.

"Well," I said thoughtfully, "there are cylinders and valves that open and close, and gas and air are brought in and mixed up and ignited by spark plugs."

"*Spark plugs,* uh huh," said another fellow.

The men stood up and began filing out of the room.

"You've got to believe me," I said, flailing my arms.

"What's there to believe," barked one of the workers angrily. "All you've told us are the names of things."

"I'm from 1985. I'm from 1985," I cried out. "I can teach you things."

"Plimpton," somebody said, "call the police. This guy is a loony. The police will know what to do with him."

And that's how it was that I met Phineas Howe. When the police tried to lock me up that first afternoon in Hartford, I demanded a trial. After a terrible scene of kicking and spouting of unknown constitutional amendments, they released me into the custody of Plimpton, who felt some strange responsibility for me. The trial was set for August 16, and Plimpton generously put me up in his house until then. Phineas Howe, without his knowledge and against his better judgment, was appointed my counsel.

A couple of days later, I met Phineas at his office to prepare our case. Phineas was about fifty years old, tall, slightly stoop-shouldered, and paunchy, with a great mop of disheveled hair. His big rubbery face had more skin than was necessary, and he looked sad, like a basset hound. He greeted me at the door with reluctance.

"You the guy from the twentieth century?" He sighed.

I nodded. He let me in. The first thing that caught my eye was the moose head on the wall. I tried to find a clear space to sit down, which wasn't easy. The couch was piled waist high with back issues of *Hunter and Field,* and the cot in the corner was spilling over with shirts and underwear. Papers and food were all over the floor. Finally I located a tiny island of space on the carpet, which belched up a huge cloud of dust when I sat down.

It was blazing hot. Phineas tossed his jacket in a random direction and rolled up his sleeves. "Now," he said, pausing to remove some wax from his ear, "tell me the facts. You realize, of course, what's at stake here. You've been charged with disturbing the peace, attempted fraud, and lunacy."

I repeated my story, while Phineas took down everything on a long yellow pad of paper. I don't think he believed one word I said, but he had been appointed by the public defender's office, and he had his job to do, and he wasn't going to lose this case for any mistake *he* made. As it turned out, Phineas had a shocking track record, but he did have a certain amount of professional pride.

We went through a series of questions and answers, with his asking the questions and my giving the answers. He asked when I was born. "December third, nineteen forty-eight," I replied. "You mean to tell me that you won't be born for sixty-eight years?" he asked in a steady voice. I stopped and figured. "Yes, that's right," I said. "I understand. I understand," Phineas said with a pained expression, and scribbled on his yellow pad. It went on like that for half an hour.

The seriousness of my situation began sinking in. "I wouldn't have been in this mess if the machinists at Colt had given me ten more minutes," I said glumly.

"Those people weren't set up for someone like you," said Phineas with an impatient brush of his hand. "I was thinking of trying to get Tom Edison down here as an expert witness. I was part of a law firm that helped him in a patent suit a few years ago. You know about Edison, don't you?"

I nodded appreciatively.

"Edison is bright enough to figure this thing out," Phineas said, and then added, "one way or the other."

After some hasty inquiries, we walked to the telegraph office and wired Menlo Park, New Jersey, where Edison worked night and day in his lab. An hour later we got a return message, which Phineas wouldn't let me read but which said something to the effect that we could go to hell and back unless I could help Edison with his lighting system for New York City. My attorney looked at me searchingly, and I said, "Certainly." This was no time to lose confidence. Phineas wired back that single word, and, within another hour, we received a second message saying that Edison would be arriving on the New York & New England line at 10:13 A.M., August 16.

The trial was held in the Court of Common Pleas, in the new brick County Building on the corner of

Trumbull and Allyn. Plimpton had wanted badly to be there with me that morning, but his daughter had suddenly become very sick with pneumonia. As I was leaving his house, I mentioned the possibility of making some penicillin, but I didn't know more than the name of it. Plimpton stared at me blankly and I left. I had grown fond of him and his wife and felt terrible about their daughter.

Phineas arrived at court looking as if he hadn't changed clothes for forty-eight hours. He was carrying several yellow pads and an armload of *Popular Science Monthly*s.

"Don't say anything except when you're up on the stand," he whispered urgently to me, "and then answer only direct questions. I'm on top of this thing." I nodded and followed him to our seats. "We're not giving those buggers one inch," he whispered again, "especially that starched son-of-a-bitch Calhoun."

"Who is Calhoun?" I whispered back. For an answer, Phineas simply glared across the room at a calm man in his late thirties, then opening a trim leather briefcase. That was Thomas Calhoun, the prosecuting attorney. He was flanked by two young assistants. All three wore immaculate gray suits. Calhoun was slender and had very black hair. Calhoun was the type of person who never utters a word that isn't right. He got his law degree from Yale. I saw all of this in the first five minutes and became depressed.

Then Judge Renshaw walked in and everyone stood up. "Renshaw doesn't have the brains to bait a fish hook with," whispered Phineas. The trial began.

Besides the parties involved, about twenty spectators had come over from the Aetna Insurance Company and sat in the back of the courtroom. Throughout the trial, they were continuously fanning themselves with paper fans advertising the Spring Grove Funeral Parlor.

I won't repeat the opening remarks. Calhoun presented the case for the town of Hartford, bringing in several men from the Armory to testify. He was brief and smooth as silk. Phineas stated our position. Judge Renshaw, a small, quiet man, seemed puzzled by the whole thing and said nothing.

Then Edison arrived. "Where's Phineas Howe?" he boomed, walking down the center aisle. The bailiff started to intercept him, but the judge raised his hand. There was a reverent hush, as everyone turned to get a glimpse of Thomas Alva Edison. Finally, a court attendant led him to our seats. Edison was a barrel-chested man with burning blue eyes. He gave me an odd look and said to Phineas, "Make this god-damned fast. I'm leaving on the twelve thirty-three."

Phineas promptly had me put on the stand and introduced Edison as his expert witness. "I intend to prove," announced Phineas, "that my client has knowledge of a technology so far advanced beyond ours that he could only be a citizen of the late twentieth century. Or beyond. This knowledge will be confirmed by the leading inventor of our age, Mr. Thomas A. Edison." The people from Aetna momentarily laid down their fans and clapped. Calhoun, to my satisfaction, shifted uncomfortably in his chair and began whispering to his aides. Phineas and I now held the trumps.

"To begin with," said Phineas, turning to me, "tell the court about your house."

"Well," I said, "I have a refrigerator, a dishwasher, a stereo, a tape deck, two telephones, a television, a video casette recorder, a microwave oven, a personal computer, and a Chrysler in the garage." Actually, I was a little embarrassed about flaunting my prosperity in front of the court like this, but Phineas had insisted. Then Phineas got me to explain what each of these items did.

"Objection," said Calhoun. "The defendant has merely invented a lot of fancy names and functions. He is wasting the court's time."

"I believe Mr. Edison will determine that," said Judge Renshaw. "Objection overruled." The judge looked expectantly at Phineas, who was rapidly paging through one of the *Popular Science Monthly*s.

Then Phineas asked me to explain to the court and to Mr. Edison how a television works.

"A radio signal comes in from a broadcast station," I began, "and is picked up by the television antenna. This signal then goes into the television and directs electricity at a picture tube, which has lots of dots on it. The dots light up when the electricity hits them. That's what makes the picture."

Edison was champing at the bit.

"Your honor," said Phineas, "will you allow Mr. Edison to question the defendant?"

Judge Renshaw nodded.

"Are there wires that go directly to this picture tube?" asked Edison.

I thought hard. "I don't think so," I answered.

"What did you say?" asked Edison.

"I don't think so," I repeated.

Edison looked as if he still hadn't heard me.

"I don't think so," I shouted.

He nodded, and said, "In that case, don't this picture tube need a vacuum inside?"

"Sounds reasonable to me." I looked over at Phineas. His hands were over his eyes.

"Does the television use direct or alternating current?" asked Edison.

I thought again. "Well, I think it comes out of the

wall alternating." There was a round of laughter in the courtroom. Phineas still had his hands over his eyes, but seemed to be peeking through his fingers.

"What's that you said?" asked Edison. He was definitely hard of hearing.

"I think it comes out of the wall alternating," I shouted.

"But is there a transformer or rectifier?" asked Edison.

"What's that?" I asked.

"A transformer increases the voltage and decreases the current, or vice versa, keeping the product constant. You lose more power with low voltage. A rectifier changes alternating current to direct current. I've been having a hell of a time with my transformers for Pearl Street. The capacitances aren't right."

What Edison was saying was extremely interesting. "Now tell me about this picture tube," he continued. "You say it lights up when electricity hits it?"

I nodded.

"How does that happen?" asked Edison. "What's this picture tube made of?"

I moved on to refrigerators. "Now a refrigerator is a marvelous machine," I said. "It keeps food cold with electricity. You can forget about hauling around big chunks of ice."

"How does one of these *refrigerators* work?" Edison asked.

"There's a motor inside," I answered loudly. "It shoves the heat outside of the refrigerator." To my surprise and embarrassment, I discovered I couldn't explain much more about refrigerators, although I was sure there wasn't much to explain.

Edison looked at his watch.

"TNT," I said firmly. "It's a very powerful explosive and excellent for weapons." The people from

Aetna stopped fanning. "Tri-nitro something," I added.

"You mean nitroglycerin?" asked Edison.

"No, not nitroglycerin. Tri-nitro something."

"What are the ingredients?" asked Edison.

"Nitrogen is one," I answered.

Edison looked at me contemptuously and said, "I don't think this fellow knows one goddamned thing about technology of any century. And he's certainly no help to me." At that, he strode out of the courtroom. Phineas, clearly shaken, told me I could sit down. Calhoun looked smug. I felt humiliated.

Judge Renshaw cleared his throat and indicated that it was time for the closing remarks. Calhoun went first.

"I believe it is clear," he said evenly, "that the defendant has demonstrated no knowledge of advanced technology, no proof that he is in fact a citizen of the twentieth century. I ask the court, therefore, to proceed on the basis that he is either a deliberate fraud, and has tried to deceive the honest people of our town, or a dangerous lunatic. The prosecution recommends five years' incarceration in Lockwood Prison, or an equal period in the Connecticut Retreat for the Insane, whichever is appropriate."

Then it was our turn. In a daring move, Phineas asked to have me put in the witness box one last time. He walked over next to me, smiled, and said quietly, "Do your friends in the twentieth century know how televisions and automobiles and computers work?"

I was keenly aware of being under oath. "There's a fellow from Acme Electronics in Cambridge who fixes my television when it's busted," I said, "but I couldn't say I really know him." I thought. "When I lived in Watertown, I knew someone who could build an automobile brake system from spare parts. Computers, well . . ." I shook my head no. "It's recommended not to take them apart."

"Do you mean to tell me," said Phineas, almost whispering, "that only a handful of people from your century understand how these things work?" Phineas was mocking me, probably to inflate his own lifeboat as the ship was sinking. I didn't blame him much, especially since he was right, but it made me mad.

"Do *you* know how a *telegraph* works?" I asked Phineas, angrily.

"I object," said Calhoun, leaping to his feet. "The knowledge and credibility of my colleague, Mr. Howe, are not relevant here. Also, it is highly improper for a defendant to argue with his own counsel."

"Objection sustained," said Judge Renshaw, yawning.

"Do *you* know how a telegraph works?" I said to the judge.

Phineas led me quickly back to our seats.

"Does the public defender have anything more to say?" said Judge Renshaw.

Phineas was writing rapidly, using a chewed-up pencil he'd whittled down to nothing with his pocket knife. "You got anything to write with?" he whispered frantically.

"Sure," I answered, and took a ballpoint pen out of my pocket. He grabbed it and continued scribbling without looking up. Suddenly he stopped and stared at the pen. He pushed the button and watched the point go in. He pushed it again and the point came back out.

"Son of a bitch," he said softly. "Will you look at this."

He got up, holding the pen, and walked over to the judge's bench. After some animated mumbling, the judge motioned for Calhoun to come over. Then the judge asked to see all of us in his chambers.

I was acquitted, of course, although the people from Aetna never knew quite why. There was a minor sensation following the trial, and a reporter from the *Hartford Times* came out. He brought a photographer with him. It seems he wanted to get a picture of me riding a horse. "The man from the twentieth century, temporarily inconvenienced, gets about by horse," or something like that.

A small crowd of people had gathered across the street from the County Building, and the reporter was there and the photographer and Phineas, and I mounted up. I guess I mounted that horse on the wrong side, because the next thing I knew I'd been pitched and was sailing by L. T. Frisie and Sons, headfirst toward a lamppost. That was the last I saw of old Hartford.

When I awoke, I was lying on the floor of my living room, covered with dust. My wife was bathing my forehead with a wet cloth. She sighed with relief as I opened my eyes. "Dear, where have you been?" she said. "I couldn't find you for over an hour, and then I heard a loud thump and found you like this, unconscious." Despite my headache, I managed a smile.

NOTES AND
REFERENCES

SMILE

The number of light particles (photons) reflected from the woman and entering the man's pupils can be calculated as follows: in diffuse daylight, the average light intensity is the same as that which hits the earth from the sun, two calories per minute per square centimeter. Using the average energy of a light photon, two electron volts, this translates to 400 thousand trillion photons per square centimeter per second. Since a pupil is about .04 square centimeters in area in bright light, this gives 30 thousand trillion photons per second *total* light entering the two pupils. Now, assuming the woman has a body area of five square feet, at a distance of twenty feet she subtends about .002 of the hemisphere viewed by the man. The fraction of incoming light she reflects is about .2. Taking this fraction of the total light gives the figure quoted.

The structure of the eye, including the dimensions of the rod and cone cells, may be gotten from chapter 13 of *Gray's Anatomy.*

Discussion of retinal molecules and their photochemical reactions may be found in "Molecular Isomers in Vision," *Scientific American,* June 1967. The number of retinal molecules hit by photons of light each second is obtained from the *total* rate of light entering the eye, as worked out in the first paragraph, above.

Discussion of the transmission of visual information to and in the brain by neurons can be found in "Brain Mechanisms of Vision," *Scientific American*, September 1979.

A discussion of how neurons work can be found in "The Neuron," *Scientific American*, September 1979.

Discussion of the optic nerve in the brain, the visual cortex and its operation can be found in "Brain Mechanisms of Vision," *Scientific American*, September 1979.

The speed of sound in air, under normal conditions, is 1100 feet per second.

The anatomy of the ear can be found in *Gray's Anatomy*, chapter 13. Chapter 11 discusses the cranial nerve.

The number of neurons in the brain can be found in "The Brain," *Scientific American*, September 1979.

E.T. CALL HARVARD

Project Sentinel is described in "Project Sentinel: Ultra-Narrowband Seti at Harvard-Smithsonian," by Paul Horowitz and John Forster, in *IAU Proceedings 112: The Search for Extraterrestrial Life* (Holland: Dordrecht, 1985). A description of the electronics and Horowitz's first search at Arecibo can be found in *Science*, August 25, 1978, p. 733.

The other full-time CETI project is that of Robert Dixon at Ohio State University. See *Icarus*, vol. 30 (1977), p. 267.

Project Ozma is described on pp. 392–393 of *Intelligent Life in the Universe*, by I. S. Shklovskii and Carl Sagan (New York: Dell, 1966).

The history of CETI conferences is described in *Acta Astronautica*, vol. 6 (1979), pp. 3–6. This particular issue of the journal has a number of articles on CETI.

The economics of CETI, favoring radio waves over rocket ships, was first pointed out by Edward Purcell, in his paper "Radioastronomy and Communication Through Space," given in 1961 and reprinted in *Interstellar Communication*, ed. A. G. W. Cameron (New York: Benjamin, 1963). The figures I give on the amount of fuel and energy required use the relativistic rocket formula in his paper and assume the United States uses about 5×10^{11} watts of power. The cost of sending a radio message also uses Purcell's estimates, but scaled to my numbers by using the fact that cost varies as the inverse square of the receiving dish size, the inverse square of the sending dish size, and the square of the distance. I've also updated the cost of electricity to ten cents a kilowatt-hour.

A description of the discovery of the first pulsar by Jocelyn Bell and Anthony Hewish at Cambridge University can be found on pp. 51–52 of *Black Holes, Quasars, and the Universe,* by Harry L. Shipman, second edition (Boston: Houghton Mifflin, 1980).

Cocconi and Morrison's paper is in *Nature,* 19 September 1959, p. 844.

Project Meta is described in "The 8-Million Channel Narrowband Analyzer," by Paul Horowitz, John Forster, and Ivan Linscott, in *IAU Proceedings 112.*

The sensitivity and range of Sentinel use the figures given in *Science,* 25 August 1978, p. 733, scaled according to the fact that range varies as the diameter of both the sending and receiving dishes.

The number of candidate stars within 1000 light years can be found on p. 195 of Purcell, *op. cit.,* scaling up the volumes and being slightly more liberal in stellar types close to our sun's.

Fontenelle's propositions regarding extraterrestrial life were made in his "Conversations on the Plurality of Worlds." An English translation of that work may be found in *The Achievement of Bernard Le Bovier de Fontenelle,* trans. L. M. Marsak (New York: Johnson Reprint Corporation, 1970). The quotes are taken from pp. 89 and 125 of "The Plurality of Worlds" in that book.

The approximate size of our galaxy and the place of our sun within it were discovered by Harlow Shapley in 1918. See the entry on him in *Dictionary of Scientific Biography* (New York: Scribner's, 1979).

The production of amino acids under primitive-earth conditions was achieved in 1953 by Harold Urey and Stanley Miller. See p. 229 of *Intelligent Life in the Universe.*

The possible discovery of another solar system, around the star Beta Pictoris, was reported in "A Circumstellar Disk Around Beta Pictoris," by B. A. Smith and R. J. Terrile, *Science,* 21 December 1984, p. 1421.

RENDEZVOUS

Information on the various missions to Halley's Comet can be found in R. Reinhard's article in *Scientific and Experimental Aspects of the Giotto Mission: Proceedings of an International Meeting on 27–28 April 1981 of the European Space Agency.* For a more up-to-date article, see "P/Halley: The Quintessential Comet," by M. J. S. Belton, *Science,* 13 December 1985, p. 1229.

Recent books on comets and their history are *The Comet Is Coming,* by Nigel Calder (New York: Penguin, 1982), and *Comet,* by Carl Sagan and Ann Druyan (New York: Random House, 1985).

Discussion of Edmund Halley's work may be found in chapter 2 of Calder, *op. cit.,* and in the entry on Halley in *Dictionary of Scientific Biography* (New York: Scribner's, 1981).

The tardiness of Halley in 1910 and the explanation may be found in chapter 5 of Calder, *op. cit.* For a fuller discussion of the nature of comets and their significance for the origin of the solar system, see "The Nature of Comets," by Fred Whipple, *Scientific American,* February 1974, and "The Spin of Comets," by Fred Whipple, *Scientific American,* March 1980. See also p. 17 of *The International Halley Watch: Report of the Science Working Group,* July 1980.

TIME FOR THE STARS

The 1982 breakup of AT&T and its possible harmful effect on the science carried out at Bell Labs is discussed in *Three Degrees Above Zero,* by Jeremy Bernstein (New York: Scribner's, 1984).

A discussion of the beginnings of science in America can be found in the first chapters of *The Physicists,* by Daniel J. Kevles (New York: Knopf, 1978).

Willard Gibbs' work in thermodynamics is described in the entry under his name in *Dictionary of Scientific Biography* (New York: Scribner's, 1979).

The quote from Millikan is from a letter to Greta Millikan on April 1, 1917, and can be found in the Robert Millikan Papers, Archives, Caltech, Box 49. It is reproduced in *The Physicists,* p. 117.

The Robert Heinlein novel is *Time for the Stars* (London: Pan Books, 1968).

Adams' comments on Madame Curie can be found in *The Education of Henry Adams,* by Henry Adams (Boston: Houghton Mifflin, 1961), chapter XXXI.

The concept of the atom was introduced by, among others, Democritus and Lucretius.

For Aristotle on circular orbits, see *On the Heavens,* trans. W. K. C. Guthrie (Cambridge: Harvard University Press, 1971), Book II, chapters III–IV.

The compounded circles were called "epicycles" and were developed around the second century B.C. by two Greek astronomers, Hipparchus and Apollonius. The most accomplished statement of this system was that of Ptolemy (100–178), in his *Almagest.* See *The Copernican Revolution,* by Thomas S. Kuhn (Cambridge: Harvard University Press, 1957), pp. 59–72.

Brahe's and Kepler's work is discussed in *The Copernican Revolution,* pp. 200–219. Kepler's laws of planetary motion were first published in his *Astronomia nova* (New Astronomy) (1609).

For Newton's acknowledgment of Kepler in his presentation of the *Principia,* see Owen Gingerich's article on Kepler in *Dictionary of Scientific Biography.*

Descartes' view of the universe as a giant clock was first presented in his "Principles of Philosophy" (1644). See *The Philosophical Works of Descartes,* vol. 1, trans. Elizabeth S. Haldane and G. R. T. Ross (New York: Dover, 1931).

A popular discussion of the Big Bang and the expansion of the universe can be found in, among other places, *The Big Bang,* by Joseph Silk (San Francisco: W. H. Freeman, 1980). This book also refers to the various observational tools needed for Hubble's discoveries.

For Aristotle's arguments for the eternity of the heavens, see his *On the Heavens,* Book I, chapters XI–XII.

Newton's arguments against bulk motion of the universe were put down in his letter to the theologian Richard Bentley on December 10, 1692. This letter is reprinted in *Theories of the Universe,* ed. M. K. Munitz (New York: The Free Press, 1965), p. 211.

For a discussion of Einstein's modification of his theory to permit a static universe, see, for example, *Gravitation,* by Charles W. Misner, Kip S. Thorne, and John A. Wheeler (San Francisco: W. H. Freeman, 1973), pp. 746–747. See also *The Big Bang,* p. 22. Einstein's original paper on cosmology appeared in *Preuss. Akad. Wiss.* (Berlin: Sitzber, 1917), pp. 142–152, and is translated in *The Principle of Relativity,* by H. A. Lorentz, A. Einstein, H. Minkowski, and H. Weyl, trans. by W. Perrett and G. B. Jeffrey (New York: Dover, 1952), pp. 175–188.

The Thoreau quote comes from *Walden,* by Henry David Thoreau, chapter 3.

Font-de-Gaume is one of the caves at Les Eyzies-de-Tayac.

FOUR FINGERS IN A HUNDRED CUBITS

Einstein's recollections of sitting in the chair in Bern are recorded in *Einstein Koen-Roku,* ed. J. Ishiwara (Tokyo: Tokyo-Tosho, 1977), and are quoted in *Subtle Is the Lord,* by Abraham Pais (New York: Oxford University Press, 1982), p. 179.

For my doctoral thesis, I tried to prove that the equal falling of bodies logically required the equivalence principle. I was able to prove this only for a limited set of conditions, and not in any generality. My partial proof appears in A. P. Lightman and David L. Lee, *Physical Review D*, 8 (1973):363.

Einstein's first paper on the equivalence principle was published in *Jahrbuch der Radioacktivitat und Elektronik*, 4 (1907):411.

De Motu, by Galileo Galilei (1592), translated in *Galileo Galilei on Motion and on Mechanics*, by I. E. Drabkin and S. Drake (Madison: University of Wisconsin Press, 1960), p. 48. See also *Gravitation*, by C. W. Misner, K. S. Thorne, and J. A. Wheeler (San Francisco: W. H. Freeman, 1973), p. 16.

R. V. Eötvös, *Math. Naturw. Ber. aus Ungarn*, 8 (1889): 65. See also *Gravitation*, pp. 16–17, 1051–1052.

Newton acknowledged the observed fact of the equal falling of bodies in gravity by simply equating the gravitational mass of a particle to its inertial mass. Einstein used the fact as a foundation for his theory.

Albert Einstein, "On the Influence of Gravitation on the Propagation of Light," *Annalen der Physik*, 35 (1911):898. English translation in *The Principle of Relativity*, by H. A. Lorentz, A. Einstein, H. Minkowski, and H. Weyl (New York: Dover, 1952), pp. 99–108. The long quote here is from p. 100 of the Dover translation.

"The happiest thought of my life" was written in a review paper for *Nature* in 1920 titled "Grundgedanken und Methoden der Relativitatstheorie in ihrer Entwicklung dargestellt." This paper was never published but has been preserved in the Pierpont Morgan Library in New York. A few paragraphs, including the above quote, can be found in *Subtle Is the Lord*, p. 178.

Eddington's measurement of the bending of light by the sun is discussed in A. P. French's chapter in *Einstein: A Centenary Volume*, ed. A. P. French (Cambridge, Mass.: Harvard University Press, 1979), pp. 98–104. It was first published in *A Report of the Relativity Theory of Gravitation,* by A. S. Eddington (London, 1920).

Page 17 of the November 10, 1919, *New York Times* is reprinted in *Understanding Relativity,* by Stanley Goldberg (Boston: Birkhauser Boston, 1984), p. 313.

Popper's measurements of 40 Eridani B were first reported in *The Astrophysical Journal,* 120 (1954):316.

Some of the Babylonian ideas about absolute space are implicit in the *Enuma elish.* See *The Babylonian Genesis,* by A. Heidel (Chicago: University of Chicago Press, 1951), and *Theories of the Universe,* ed. M. K. Munitz (New York: The Free Press, 1957), pp. 9, 12.

On the Heavens, by Aristotle, trans. W. K. C. Guthrie, Loeb Classical Library, Aristotle, vol. 6 (Cambridge, Mass.: Harvard University Press, 1971). The sections I quote from can also be found in *Theories of the Universe,* pp. 93, 96–97.

Principia, by Isaac Newton, trans. in *Principia* (Berkeley: University of California Press, 1962). I quote from pp. 6 and 8 of vol. 1 of the Berkeley edition.

Albert Einstein, "The Foundation of the General Theory of Relativity," *Annalen der Physik,* 49 (1916), trans. in *The Principle of Relativity,* pp. 111–164. I quote from p. 113 of the latter.

"This assumption of exact physical equivalence makes it impossible . . . to speak of the absolute acceleration . . ." appears in *The Principle of Relativity,* p. 100.

IN HIS IMAGE

A good book on the older history of the belief in extraterrestrial life is Steven J. Dick, *A Plurality of Worlds* (Cambridge: Cambridge University Press, 1982).

The Search for Extraterrestrial Intelligence, ed. Philip Morrison, John Billingham, and John Wolfe (Washington: NASA, 1977).

The Summa Theologica of St. Thomas Aquinas, trans. Fathers of the English Dominican Providence (London, 1921), pp. 261–262. See also *A Plurality of Worlds,* p. 27.

For a discussion of Greek atomism and of Aristotle's cosmological theories, see *A Plurality of Worlds,* chapter 1. A fuller discussion of Aristotle's physics and cosmology is given in D. J. Allan, *The Philosophy of Aristotle* (London, 1952). See also Aristotle's *On the Heavens,* trans. W. K. C. Guthrie, Loeb Classical Library (Cambridge, Mass.: Harvard University Press, 1971), Book I, chapter VIII.

St. Thomas' synthesis of Christianity and Aristotelian science is discussed in the entry on Aquinas in *Dictionary of Scientific Biography* (New York: Scribner's, 1981).

For the 1277 statement supporting the plurality of worlds by the Bishop of Paris, Etienne Tempier, see *Medieval Political Philosophy: A Source Book,* ed. Ralph Lerner and Muhsin Mahdi (Glencoe, N.Y., 1963), pp. 337–354. The argument of the bishop was called the "principle of plentitude."

Kepler's and Galileo's contributions to the plurality of worlds debate are discussed in *A Plurality of Worlds,* chapter 4. See pp. 72–73 for Kepler's discussion of the new star.

Galileo's hypotheses concerning lunar atmospheres and oceans are discussed in *Discoveries and Opinions of Galileo,* ed. Stillman Drake (New York, 1957), pp. 31, 34, 39–40.

Kepler's conclusion about artificial construction on the moon is in *Kepler's Somnium: The Dream or Posthumous Work on Lunar Astronomy,* trans. Edward Rosen (Madison: University of Wisconsin Press, 1967), p. 160.

Kepler's work on the brightness of stars is in *Kepler's Conversation with Galileo's Sidereal Messenger,* trans. Edward Rosen (New York and London, 1965), pp. 34–36.

Kepler's quote on the nobility of man comes from *Kepler's Conversation,* p. 43.

Newton's correspondence with Bentley is in *Isaac Newton's Papers and Letters on Natural Philosophy,* ed. I. B. Cohen (Cambridge, Mass.: Harvard University Press, 1958). Some of Newton's comments regarding Divine intervention in the formation of the solar system are in the General Scholium of *The Principia.* See for example *The Principia,* by Sir Isaac Newton, revised by Florian Cajori (Berkeley: The University of California Press, 1962), pp. 543–547.

M. A. Hoskins, "The Cosmology of Thomas Wright of

Durham," *Journal for the History of Astronomy,* 1 (1970): 44–52. See also *A Plurality of Worlds,* p. 159.

For a discussion of natural theology, see John Dillenberger, *Protestant Thought and Natural Science* (New York, 1960). For the cultural offshoots, such as the works of Rousseau, Coleridge, Wordsworth, Turner, and Constable, see *Civilization,* by Kenneth Clark (New York: Harper and Row, 1969), chapter 11.

John Wilkins, *The Discovery of a World in the Moone* (1638), facsimile reprint (Delmar, 1973), proposition 2.

Bentley's comments are in *Isaac Newton's Papers and Letters,* pp. 356–358.

Milton's *Paradise Lost,* Book VIII. See, for example, *The Harvard Classics,* ed. Charles W. Eliot (New York: Collier and Son, 1909), vol. 4, pp. 248 and 250.

Descartes' "Principles of Philosophy," in *The Philosophical Works of Descartes,* trans. Elizabeth S. Haldane and G. R. T. Ross (New York: Dover, 1931), vol. 1.

Fontenelle's "A Plurality of Worlds," in *The Achievement of Bernard Le Bovier de Fontenelle,* trans. Leonard M. Marsak (New York: Johnson Reprint Corporation, 1970). Details of Fontenelle's life are given in the introduction to this book, by Marsak.

The final triumph of the concept of the plurality of worlds is discussed in *A Plurality of Worlds,* chapter 6.

A TELEGRAM FROM CLARENCE

A principal source on the Scopes trial is the book *Six Days or Forever?,* by Ray Ginger (Boston: Beacon Press, 1958), which uses as sources original transcripts, newspaper articles, and eye-witness reports. The scientists who made statements and summaries of their statements can be found in chapter 8.

A summary of Shapley's work may be found in *Dictionary of Scientific Biography* (New York: Scribner's 1981).

The quote from Shapley comes from his autobiography,

Through Rugged Ways to the Stars (New York: Scribner's, 1969), pp. 59–60.

Bruno's concept of stars as suns can be found in *Theories of the Universe,* ed. M. K. Munitz (New York: The Free Press, 1957), p. 183.

Newton estimated the distance to nearby stars in his *System of the World.* See *Sir Isaac Newton's Principles* (Berkeley: University of California Press, 1962), p. 596.

The first measurements of the distances to stars, in 1838, are mentioned in *Exploration of the Universe,* by George Abell (New York: Holt, Rinehart and Winston, 1969), second edition, p. 375. Galileo's astronomical work is mentioned on p. 50. Herschel's work is mentioned on p. 454.

A watt is 10^7 erg/s. The sun has a luminosity of 4×10^{33} erg/s.

Kapteyn's work is discussed in *Dictionary of Scientific Biography.*

Shapley's work on ants is discussed in *Through Rugged Ways to the Stars,* pp. 65–66. The rugged hike up the mountain is mentioned on p. 51.

Globular clusters are discussed in *Exploration of the Universe,* pp. 142–143. Cepheids are discussed on pp. 481–485.

Leavitt's, Bailey's, and Shapley's work with Cepheids is summarized in the entry under Shapley in *Dictionary of Scientific Biography.*

Shapley is shown sitting at his "famous rotating desk" in one of the photographs following p. 52 in *Through Rugged Ways to the Stars.*

THE ORIGIN OF THE UNIVERSE

A good biographical article on Stephen Hawking is "The Universe and Dr. Hawking," by Michael Harwood, *The New York Times Magazine,* January 23, 1983, p. 16ff.

Hawking's work on the initial condition of the universe was first presented in preliminary form at a study week in 1981 sponsored by the Pontifical Academy of Sciences in

Rome. This was published in *Astrophysical Cosmology: Pontificiae Academiae Scientarium Scripta Varia,* 48 (1982):563. Subsequent papers have included J. B. Hartle and S. W. Hawking, *Physical Review,* D28 (1983):2960, and S. W. Hawking, *Nuclear Physics,* B239 (1984):257.

The discovery of the nonuniform flow of time and the behavior of subatomic matter refer to the theory of special relativity and to quantum mechanics, respectively.

The unified theory of electromagnetic and weak nuclear forces, developed by Sheldon Glashow, Steven Weinberg, and Abdus Salam in the 1960s, predicted the existence of the W and Z particles. In 1983, a team led by Carlo Rubbia discovered these particles with the CERN accelerator near Geneva.

Neutron stars, consisting of almost pure neutrons and measuring about fifteen miles in diameter, were predicted in the early 1930s by Lev Landau and by Fritz Zwicky. The first one of these stars, a pulsar, was discovered by Jocelyn Bell and Anthony Hewish in 1967.

Einstein's prediction about the deflection of starlight was first confirmed in 1919 by a team led by Arthur Eddington. See, for example, the chapter by A. P. French in *Einstein: A Centenary Volume,* edited by A. P. French (Cambridge: Harvard University Press, 1979). The student who asked Einstein what he would have done if his theory had been refuted and who recorded his reply was Ilse Rosenthal-Schneider. See Gerald Holton's article in *Daedalus,* Spring 1968, p. 653; and also *Einstein: The Life and Times,* by Ronald W. Clark (New York: World Publishing, 1971), p. 230.

The Golden Bough, by James G. Frazer (New York: Avenel Books, 1981) (originally published in 1890 in two volumes as *The Golden Bough: A Study in Comparative Religion*). The two specific examples given, of making rain and making wind, are given on pp. 13 and 26, respectively.

The Bishop of Paris, Etienne Tempier, issued the so-called Condemnation of 1277 on March 7, 1277. For some of the condemned propositions bearing on the power of God, see *History of Christian Philosophy in the Middle Ages,* by Etienne Gilson (New York: Random House, 1955), pp.

405–408. See also *The Masks of God: Creative Mythology,* by Joseph Campbell (New York: Viking, 1968), p. 400.

The quote from Milton's *Paradise Lost* comes from Book VIII. See, for example, *The Harvard Classics,* vol. 4 (New York: Collier and Son, 1909), p. 248.

For the quotes from Newton, see, for example, *The Principia* (trans. Andrew Motte in 1729), rev. and ed. Florian Cajori (Berkeley: University of California Press, 1962), vol. II, pp. 544–545.

Among the scientists and mathematicians who continued to debate the stability of the solar system were Pierre Simon de Laplace, Karl Theodor Weierstrass, and Jules Henri Poincaré.

Leading figures who believed in gradual transformations, the so-called Uniformitarian school, were James Hutton (1726–1797) and Charles Lyell (1797–1875). A leading figure who believed in sudden catastrophes, according to God's intervention, was Georges Cuvier (1769–1832).

The quote from Henry Adams comes from *The Education of Henry Adams,* by Henry Adams (first published in 1907) (Boston: Houghton Mifflin, 1961), p. 451.

The quote from Hawking comes from "The Quantum State of the Universe," by Stephen Hawking, *Nuclear Physics,* B239 (1984):258.

TINY PATTERNS

A description of Nakaya's experiment, as well as interesting general information on snow, can be found in *The Wonder of Snow,* by Corydon Bell (New York: Hill and Wang, 1957).

Kepler's work on snow is mentioned in *Dictionary of Scientific Biography* (New York: Scribner's, 1979).

Types of crystal structures found in solids are discussed in the first chapter of *Introduction to Solid State Physics,* by Charles Kittel (New York: Wiley, 1967). The two ways of packing golf balls are called "hexagonal close-packed structure" in Kittel's book, p. 25.

A typical snowflake is about a tenth to a hundredth as thick as it is wide. Using this, one can calculate how many water molecules are in a snowflake of any size. A large snowflake can weigh a hundred-thousandth of a gram.

The Origin of Species (1859), by Charles Darwin.

Bentley's work is described in *The Wonder of Snow*, as well as in many other places.

The major types of snow may be found in numerous references, including *The Wonder of Snow*.

The relationship of snow crystal types to weather conditions is discussed in "Snow Crystals," by Charles and Nancy Knight, *Scientific American*, January 1973, p. 100.

Langer's work is described in *Reviews of Modern Physics*, vol. 52 (1980), p. 1, and in *Physical Review A*, vol. 29 (1984), p. 330.

The last part of the last sentence of the essay, "the only other sound was 'the sweep of easy wind and downy flake,'" is a slight paraphrase from the last two lines of Robert Frost's poem "Stopping by Woods on a Snowy Evening."

TO CLEAVE AN ATOM

Civil defense information can be found in "Forty Years of Civil Defense," by Allan M. Winkler, in *Bulletin of the Atomic Scientists*, June/July 1984, p. 16.

A good biography of Fermi is *Atoms in the Family*, by Laura Fermi (Chicago: University of Chicago Press, 1954). Fermi's account of the first chain reaction is in *American Journal of Physics*, December 1952, p. 536.

General references on the history of fission are: *The Physicists*, by C. P. Snow (Boston: Little, Brown, 1981); *Matter*, by Ralph E. Lapp and the editors of Time-Life Books (New York: Time-Life Books, 1963), chapter 8; *The Discovery of Nuclear Fission: A Documentary History*, by Hans G. Graetzer and David L. Anderson (New York: Van Nostrand Reinhold, 1971); "The Discovery of Fission," by Otto R. Frisch and John A. Wheeler, *Physics Today*, November 1967, pp. 43–52.

Biographical information on Frederick Soddy, as well as other scientists referred to, may be found in *Dictionary of Scientific Biography* (New York: Scribner's, 1974). Soddy's first comment on the latent energy of the atom was "The Disintegration Theory of Radioactivity," in the *Times Literary Supplement*, June 26, 1903, p. 201; his 1906 statement was in "The Internal Energy of Elements," in *Journal of the Proceedings of the Institution of Electrical Engineers*, Glasgow, vol. 47 (1906), p. 7.

The World Set Free, by H. G. Wells (New York: E. P. Dutton, 1914). Rutherford and Soddy are mentioned by name on p. 40. The "atomic bomb" appears on p. 109.

Frisch's quote referring to his visit with Meitner is in the article "The Discovery of Fission," p. 47.

Frisch and Meitner's original paper discussing the Hahn-Strassman results in terms of Bohr's liquid-drop model is "Disintegration of Uranium by Neutrons: A New Type of Nuclear Reaction," in *Nature*, vol. 143 (1938), p. 239. My explanation differs somewhat from theirs but follows from the same underlying physics.

Bohr's comments in his conversation with Frisch can be found in "The Discovery of Fission," p. 47.

Einstein's letter to Roosevelt can be found in many places. One is *The Physicists*, p. 178.

Most chemical reactions release about 1 kilocalorie per gram. Fission of uranium releases 200 MeV per uranium atom, which translates to about 20 million kilocalories per gram.

Historical information on nuclear power plants can be found in encyclopedias under "Nuclear Power." The statistics on the state of such plants in 1984 can be found in *Time*, February 13, 1984, p. 34.

Information on the defense against the V-1 buzz bombs is in *Living with Nuclear Weapons*, by the Harvard Nuclear Study Group: Albert Carnesale, Paul Doty, Stanley Hoffmann, Samuel P. Huntington, Joseph S. Nye, Jr., and Scott D. Sagan (New York: Bantam, 1983), p. 31.

Information on current nuclear arsenals can be found in *Living with Nuclear Weapons*.

The study by Educators for Social Responsibility was reported at a conference in October 1982 and can be found in "Children and Nuclear War," by John E. Mack, *IPPNW Report,* vol. 2, no. 1, Winter 1984.

The word *stuck* was inspired by Freeman Dyson's use of the word in *Weapons and Hope* (New York: Harper and Row, 1984), chapter 17.

LOST IN SPACE

The best recent estimate of the logistics and cost of a space-based ballistic missile defense system is "Cost of a Space-Based Laser Ballistic Missile Defense," by G. Field and D. Spergel, *Science,* March 21, 1986, p. 1387.

Lawrence West was quoted in William Broad's article "The Young Physicists: Atoms and Patriotism Amid the Coke Bottles," *The New York Times,* 31 January 1984, Section C, p. 1.

Robert Oppenheimer's quote comes from *In the Matter of J. Robert Oppenheimer: U.S.A.E.C. Transcript of Hearings Before Personnel Security Board* (Washington, D.C.: U.S. Government Printing Office, 1954), p. 81.

THE DARK NIGHT SKY

Edmund Halley, "Of the Infinity of the Sphere of Fix'd Stars," *Philosophical Transactions,* xxxi (1720–1721): 22–24. This was read on 9 March 1721. Requoted in *Stellar Astronomy,* by Michael Hoskin (Cambridge, England: Science History Publications, 1982), p. 96

A full account of Olbers' paradox is given in *The Paradox of Olbers' Paradox,* by S. L. Jaki (New York: Herder and Herder, 1969).

The quote from Digges can be found in *From the Closed World to the Infinite Universe,* by Alexander Koyre (Baltimore: John Hopkins University Press, 1974), p. 36.

Aristotle's cosmology, including the concept of a spatially

finite and eternal universe, is given in *On the Heavens,* trans. W. K. C. Guthrie (Cambridge, Mass.: Harvard University Press, 1971). Relevant sections can be found in *Theories of the Universe,* ed. M. K. Munitz (New York: The Free Press, 1965), pp. 89–100.

The quote from Bruno can be found in *From the Closed World to the Infinite Universe,* p. 45.

Moses Maimonides' refutation of Aristotle's cosmology is in his *Guide of the Perplexed,* published in 1190. A translation is by M. Friedlander (New York: Dover, 1956).

Eddington's first announcement of subatomic energy as the power source of stars was in *Report of the Eighty-eighth Meeting of the British Association for the Advancement of Science* (1920), p. 34.

LET THERE BE LIGHT

Dr. Geller's colleagues in the discovery of the bubblelike structures were John Huchra and Valerie de Lapperant. Their results were first announced at a meeting of the American Astronomical Society in Houston on January 5, 1986, and published in *The Astrophysical Journal Letters,* 302 (1986):L1.

Grimaldi's work in optics was published under the title *De Lumine* in 1665. For a description of Grimaldi's life and work, see *Dictionary of Scientific Biography* (New York: Scribner's, 1981) and *The Nature of Light,* by Vasco Ronchi, trans. V. Barocas (Cambridge, Mass.: Harvard University Press, 1970), pp. 124–149.

For a description of Lenard's life and work see *Dictionary of Scientific Biography.* Almost all of Lenard's scientific papers have remained in German. His famous paper on the discovery of the photoelectric effect is in *Annalen der Physik,* 8 (1902): 149. Einstein's explanation of the photoelectric effect in terms of photons, for which he won the Nobel Prize, was published in *Annalen der Physik,* 17 (1905):132.

Richard Feynman describes his ability to imagine invisible angels and his inability to imagine light waves in *The*

Feynman Lectures on Physics, by Richard P. Feynman, Robert B. Leighton, and Matthew Sands (New York: Addison-Wesley, 1964), vol. II, pp. 20–9.

An excellent history of Greek and Roman ideas about light and vision is given in *The Nature of Light,* chapter 1.

For an English translation of *De Rerum Natura* side by side with the original Latin, see *Lucretius: De Rerum Natura,* trans. W. H. D. Rouse and M. F. Smith, in the Loeb Classical Library (Cambridge, Mass.: Harvard University Press, 1982). The relation between sense perceptions and the properties of atoms is in 2.398–408. The comparison between light and rain is in 2.388. The quote beginning "Bright objects . . ." comes from 4.324–331.

Ibn al-Haytham is also known as Alhazen. His great work on light is titled *De Aspectibus* in the Latin translation. His life and work are discussed in *Dictionary of Scientific Biography,* and in *The Nature of Light,* pp. 45–59.

A good biography of William Herschel is *William Herschel,* by J. B. Sidgwick (London, 1953), and a discussion of his work is in *William Herschel: Pioneer of Sidereal Astronomy,* by Michael A. Hoskins (London, 1959). See also *The Nature of Light,* p. 274, for a brief discussion of Herschel's measurements of the colors in sunlight.

Einstein's comment on Maxwell appears in *James Clerk Maxwell* (Cambridge: Cambridge University Press, 1931), p. 66, and is reprinted in *Subtle Is the Lord,* by Abraham Pais (New York: Oxford University Press, 1982), p. 319.

There are many biographies on Maxwell. A good place to start is *Dictionary of Scientific Biography.* Maxwell's original papers have been reprinted in *The Scientific Papers of James Clerk Maxwell,* ed. W. D. Niven, 2 vols. (New York: Dover). The paper quoted from is "A Dynamical Theory of the Electromagnetic Field" (London: Royal Society Transactions, vol. CLV, 1864) and appears in volume 1 of Maxwell's *Scientific Papers.*

For a discussion of Roemer's discovery of the speed of light, see "Roemer and the First Determination of the Velocity of Light," by I. B. Cohen, in *Isis*, 31 (1940): 327. A review of Roemer's life and work is given in *Dictionary of Scientific Biography*.

Maxwell's article on "Ether" is reprinted in *The Scientific Papers of James Clerk Maxwell*, vol. 1. The quote given appears on p. 763. See also *Subtle Is the Lord*, p. 111. A good book on ether is *The Ethereal Ether*, by L. S. Swenson (Austin: University of Texas Press, 1972).

The Michelson-Morley experiment has been discussed widely in the popular and technical literature. See, for example, Silvio Bergia's chapter in *Einstein: A Centenary Volume*, ed. A. P. French (Cambridge, Mass.: Harvard University Press, 1979); *Michelson and the Speed of Light*, by Bernard Jaffe, Science Study series (Garden City, N.Y.: Doubleday, 1960); and the entry on Michelson in *Dictionary of Scientific Biography*.

The direct influence of the Michelson-Morley experiment on Einstein is controversial. It is likely that Einstein's dismissal of the ether, and of the absolute frame of rest that it represented, was motivated more by much older experiments with electricity and magnetism. Gerald Holton has given an extensive discussion of these influences in his book *Thematic Origins of Scientific Thought* (Cambridge, Mass.: Harvard University Press, 1973), pp. 261–352. (Reprinted from *Isis*, 60 [1969]: 133.)

Einstein's celebrated paper on special relativity, in which he postulates that the speed of light is independent of the motion of the observer, first appeared in *Annalen der Physik*, 17 (1905), and is reprinted under the title "The Electrodynamics of Moving Bodies," in *The Principle of Relativity*, by H. A. Lorentz, A. Einstein, H. J. Minkowski, and H. Weyl, trans. W. Perrett and G. B. Jeffrey (New York: Dover, 1952).